BREASTS

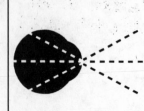

BREASTS

A NATURAL AND UNNATURAL HISTORY

FLORENCE WILLIAMS

THORNDIKE PRESS

A part of Gale, Cengage Learning

GALE
CENGAGE Learning®

Detroit • New York • San Francisco • New Haven, Conn • Waterville, Maine • London

LIBRARY OF CONGRESS CATALOGING-IN-PUBLICATION DATA

Williams, Florence, 1967–
    Breasts : a natural and unnatural history / by Florence Williams. — Large
print edition.
      pages cm. — (Thorndike Press large print nonfiction)
    Includes bibliographical references.
    ISBN 978-1-4104-5211-5 (hardcover) — ISBN 1-4104-5211-5 (hardcover)
    1. Breast—History. 2. Breast—Psychological aspects. 3. Large type books.
    I. Title.
    QM495.W55 2012b
    612.6'64—dc23
                           2012024854

Published in 2012 by arrangement with W. W. Norton & Company, Inc.

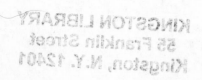

Printed in the United States of America
1 2 3 4 5 6 7 16 15 14 13 12

BREASTS

In memory of my grandmothers,
Florence Higinbotham Williams
and Carolyn Loeb Boasberg,
and my mother,
Elizabeth Friar Williams.

CONTENTS

9

Introduction: Planet Breast

> Save the Tatas
> — BUMPER STICKER

Funbags. Boobsters. Chumbawumbas. Dingle bobbers. Dairy pillows. Jellybonkers. Num nums. When I was growing up, my mother called them *ninnies.* That word, according to *Webster's,* means "fools," and lists *nitwits, nutcases,* and *boobs* as synonyms. For my own children, I amended the word to *nummies,* thinking it a bit kinder. Looking up its etymology recently, I found the word defined as "yummy," but its origin may stem from *numbskull.* We love breasts, yet we can't quite take them seriously. We name them affectionately, but with a hint of insult. Breasts embarrass us. They're unpredictable. They're goofy. They can turn both babies and grown men into lunkheads.

For such an enormously popular feature of the human race — even today, when they

11

are bikinied, bared, flaunted, measured, inflated, sexted, YouTubed, suckled, pierced, tattooed, tassled, and in every way fetishized — it's remarkable how little we actually know about their basic biology. We know some things: they appear out of nowhere at puberty, they get bigger in pregnancy, they're capable of producing prodigious amounts of milk, and sometimes they get sick. We know men even get them once in a while, and that tweaks us out.

Not even the experts among us are certain why all these things happen, or even why we have breasts in the first place. But the urgency to know and understand breasts has never been greater. Modern life has helped many of us live longer and more comfortably. It has also, however, taken a strange and confounding toll on our breasts. For one thing, they are bigger than ever, according to lingerie manufacturers and purveyors who are ever increasing their cup offerings to sizes like H and KK. We are sprouting them at younger and younger ages. We are filling them with saline and silicone and transplanted stem cells to change their shape. Most of us are not using them to nurture infants anymore, but when we do, our breast milk contains industrial additives never tasted by our ancestors and never meant to be digested by humans at all. More tumors form in the breast than in any other organ, making breast cancer the most common malignancy in women worldwide. Its incidence has almost doubled since the 1940s and is still rising. Breasts are living a life they've never lived before.

Fortunately, scientists are beginning to unveil the secrets of breasts, and with those secrets, a new way of looking at human health and our decidedly complicated place in nature. To understand the transformation, we need to go back in time, to the very beginning. We must first ask, Why breasts? Why us? We share 98 percent of our genes with chimps, but among that immeasurable 2 percent are the ones governing breasts. Chimps, unlucky sods, don't have them. In fact, we are the only primates so endowed with soft orbs from puberty onward. Other female primates develop small swellings while lactating, but they deflate after weaning. Breasts are a defining trait of humanity, and mammary glands define our entire taxonomic class. Carolus Linnaeus understood. That's why he named us mammals.

Breasts are us.

I didn't think a lot about my breasts until I

became a mother. My breasts developed about the normal time. I liked them fine. They were small enough not to get in the way of sports or cause backaches, big enough that I knew they existed, symmetrical enough to look okay in a bathing suit on the rare occasions when I wore one while growing up in New York City. I wasn't like Nora Ephron, who wrote an essay for *Esquire* about how she obsessed over her small breasts in the 1950s, the era of the torpedo bra, in California: "I would sit in the bathtub and look down at my breasts and know that any day now, any second now, they would start growing like everyone else's. They didn't."

Poor Nora. Her worry acknowledged a truth that had been evolving since the sun set on the Pleistocene: breasts are really important. Consider this: in our mammalian ascent, being breastfed allowed us a youthful pass from gathering, chewing, digesting, and purifying food found in nature. Other animals such as reptiles had to live near specialized, high-fat food sources. Mammals just had to be near their moms, who do all that work for them. Mammals had more flexibility during times of climate change and food scarcity. After lactation evolved (from sweat glands) in the Mesozoic, mammals gained ascendancy over dinosaurs. The world became a different place.

Breasts have helped advance our own species's evolution in ways both self-evident and unexpected. With their rich stores of milk, they allowed our newborns to be born smaller and our brains to grow bigger. Having smaller babies meant our hips could be smaller, assisting our ascent into bipedalism. Breast-feeding may well have enabled the development of gesture, intimacy, communication, and socialization. Our nipples helped develop and prepare the human palate for speech and gave us a reason to have lips. So, in addition to greasing the way to our global domination, breasts begat the fine art of kissing. It was a tall order, but breasts were up to it.

Millions of years of evolution and environmental pressure created an organ that was pretty darn fabulous, or so we thought.

Mine looked fabulous for about nine months, while I was pregnant with my first child. After he was born, my breasts became wondrously utilitarian for the first time. But for a piece of finely tuned evolutionary machinery, mine often malfunctioned. They became objects of betrayal, frustration, self-doubt, and excruciating pain. Metaphors of aeronautics were now applied with disturbing frequency. I didn't employ the correct "latch-on" or "suction disengagement," and my nipples paid the price. A week after my son was born, I came down with my first case of mastitis, a practically medieval systemic infection that starts in a clogged milk duct. I would endure three more cases before the first year was out.

Although I grew to love breast-feeding, I am not a dewy-eyed sprite about it. Breasts are virtually the only organ the body has to learn how to use. The process isn't for everyone. I was certainly swayed by visions of the purity and goodness of breast milk. While baby formula is derived from either cow's milk or soy protein, human breast milk is perfectly suited to the human baby, as we are often told in the mommy literature. It contains hundreds of substances — including ones that fight germs — many of which cannot or have not been synthesized in formula. Breast milk is always the right temperature; it has the correct balance of lipids, proteins, and sugars. It is medicinal, nutritious, and, to a baby, delicious. It was designed to be the perfect food, and I, new mother, was sold.

I was happily nursing my second child, blithely backstroking through that magic bubble known as the mother-infant pair-bond, when I stumbled upon a news report that would forever alter my perception of breasts. I read that scientists were finding industrial chemicals in the tissues of land and marine mammals as well as in human breast milk. So much for the blissful insularity of early motherhood. Along with their more exalted roles, breasts, I learned, are also the catchment for our environmental trespasses. I realized my breasts connected me not just to my children, but me (and by extension, my children) to my neighboring ecosystem. Breastfeeding, it turns out, is a very efficient way to transfer our society's industrial flotsam to the next generation.

I released my breast from my daughter's airlock and searched for answers. What toxic load had I already bequeathed my children by nursing them? What did it mean to their health, and to mine? Was it still okay to breast-feed? How did these chemicals interfere with our bodies? Could we ever make our milk pure again?

I did what journalists do and wrote about it. For a piece published in the *New York Times Magazine,* I sent my breast milk off to Germany to have it tested for flame-retardants, a common class of chemicals known to accumulate in fat and cause health problems in lab animals. My levels came back higher than I expected and ten to a hundred times higher than those found in European women. My exposure came from electronics, furnishings, and food. I also had my breast milk tested for other chemicals, including perchlorate, a jet-fuel ingredient, which certainly is not what baby has in mind for dinner. My results kept coming back positive, with levels about "average" for Americans. It was a discouraging revelation of how thoroughly polluted we've become in the early twenty-first century.

"Well, at least your breasts won't spontaneously ignite!" quipped my husband, trying to make the best of a situation over which we were virtually helpless. But I was reeling. The chemical cocktail in my chest collided with the journalist in my head. I wanted to find out how this elixir of evolution had come to meet such a diminished fate. Beyond that, I had to wonder how modern life was changing our breasts in other ways, and changing our health.

The answers weren't always easily forthcoming.

Perhaps not surprisingly, breasts have often eluded clearheaded thinking. Every set of eyes sees them a little bit differently. Rather than naming us all mammals, Linnaeus could have classified us by our shared ear-bone construction or four-chambered heart, but in singling out our unique mammae, he appears to have had a political motive as well as a scientific one. Linnaeus was the father of seven children. One of the conventions he abhorred was the practice of wet-nursing, in which the infants of the European middle and upper classes were literally farmed out to be breast-fed by surrogates. As a result, many babies died of malnutrition and disease. In 1752, a few years before Linnaeus introduced the term *Mammalia* into his tenth edition of *Systema Naturae,* he wrote a treatise on "Mercenary Wet-Nurses." The science historian Londa Schiebinger argues that while Linnaeus cared about infant health, he was also deeply perturbed by the possibility of greater equality between the sexes during the Enlightenment. To Linnaeus, a woman's rightful place was in the home, acting as nature intended. To prove it, we would now be called mammals.

Then again, maybe Linnaeus just liked breasts. He was hardly the only man of science to conscript this body part into ideological service. The breast has always been a favorite of evolutionary biologists, who imbue it with colorful origin stories that may or may not be rooted in reality. Scientists have spent decades looking (and looking) at the breast, trying to figure out how on earth humans got so lucky. For years now, many have been seeing breasts as a wonderful adornment — like the peacock's tail — designed to attract the opposite sex. When humorist Dave Barry wrote, "The primary biological function of breasts is to make males stupid," he was summing up a half century of scholarship on the subject. Breasts, said a whole generation of academics, evolved because men loved them and preferred to mate with early cave women fortunate enough to have them.

By the latter quarter of the twentieth century, though, as women climbed the ranks of anthropology and biology departments, they had — and continue to have — some other ideas about how these mysteries arrived on the female chest. These gatecrashers hypothesized that it was actually the maternal woman who drove the evolution of breasts. Perhaps our she-ancestors needed those few extra grams of thoracic fat to gestate and nurture their babies, who are, after all, the pudgiest little primates in the history of the earth.

The debate over breast evolution is important, because the creation stories color how we see breasts, how we use them, and how we burden them with our expectations. Because the dominant story has been all about the visuals, it discounts what's actually *in* breasts. How do they work? How are they connected to the rest of the body, and how are they affected by a larger ecology?

I wasn't expecting to ponder these questions. But writing that magazine article opened up a whole new world of environmental health. Our bodies, I learned, are not temples. They are more like trees. Our membranes are permeable; they transport both the good and the bad from the world around us. Twentieth-century medicine taught us that germs make us sick, but human health, I came to realize, is far more complex than this model. It is also governed by the very places we live and the small-print ingredients in the water we drink, by the molecules we touch and breathe and ingest every day. It became increasingly clear that we are not simply agents of environmental change; we are also objects of that change.

And breasts are a particularly vulnerable and visible pair of objects. To their credit and their detriment, breasts were built to be great communicators. From their earliest, circular beginnings, breasts have been highly sensitive to the world around them, conversing both within and outside the body. Because breasts store fat, they store toxic, fat-loving chemicals. Some of these substances persist for decades in our tissues. Breasts also contain a dense supply of receptors that sit on cell walls like hungry Venus flytraps, waiting around to catch passing molecules of estrogen, nature's very first hormone. It's an ancient habit. Before advanced organisms produced their own estrogen, cells had to get it from elsewhere. Our twenty-first-century breasts are still looking for it, and they're getting a lot more than they bargained for. Plants make estrogenic compounds, and so do chemical and pharmaceutical corporations, often unintentionally. These chemicals — estrogen variants or mimics — interact with our cells in ways that are both subtle and overt. Our breasts soak up pollution like a pair of soft sponges.

To understand why our breasts so easily consort with molecules of bad repute, I needed to learn how cells work and how they respond to changes in the environment. During a year as a research fellow in environmental journalism and then as a visiting scholar at the University of Colorado, I studied cells, genetics, and endocrinology. My continued quest took me to corners both dark and well lit, to the experts working in the emerging fields of epigenetics and environmental endocrinology, as well as in the more established fields of evolutionary, cellular, and cancer biology.

What I found was not only unsettling and profound, but also at times funny and provocative. Take, for example, the Barbie debate. Women with more hourglass waist-to-hip-to-breast ratios produce slightly higher levels of estrogen as a general rule. Sounds desirable, right? But those women may also be more likely to cheat on their mates and to get breast cancer. In fact, women with fewer curves can hold their own just fine, thank you, as some indignant researchers have pointed out. In times of trouble and stress, it may be these women — with their slightly higher levels of so-called male hormones — who can bring home the mastodon and slap competitors upside the head. That's pretty sexy. (There's an interesting male corollary to this: men with bigger muscles attract more mates but appear to have weaker immune systems. Beauty comes at a price.)

I learned that breast milk, once the magic mojo of evolution, might now actually be devolving us, holding back our potential. Toxins in breast milk have been associated with lower IQ, compromised immunity, behavioral problems, and cancer. Our modern world is not just contaminating our breast milk. It is also reshaping our children, contributing to earlier puberty in girls. Breasts are often one of the first signs of sexual development. When girls sprout breasts earlier, they face an increased risk of breast cancer later, for reasons I will explore. In fact, at every life stage of the breast, from infancy to puberty to pregnancy to breast-feeding to menopause, our modern environment has left a mark.

As civilization marches on, we have also steered breasts away from their natural lives by hiring wet nurses, entering nunneries, controlling our reproduction, and seeking to alter the breast cosmetically. After her mastectomy in the early 1970s, my grandmother wore a breast "prosthesis" that bore the silhouette and heft of a nuclear weapon head. Ironically, these devices were promoted — and later designed — by none other than the inventor of Barbie, Ruth Handler, herself a breast cancer patient. Today's fake and enhanced breasts are far more naturalistic. Nearly everyone wants a piece. Wonderbra sales in the United States top $70 million annually.

In countless ways, modernity has been good for women, but it hasn't always been so good for our breasts. The global rise in the incidence of breast cancer is partly influenced by better diagnostics and an aging population. But those factors aren't enough to explain it. The wealthiest industrialized countries have the highest rates of breast cancer in the world. Family history accounts for only about 10 percent of breast cancers. Most women (and increasingly, men) who get the disease are the first in their families to do so. Something else is going on, and that something else is linked to modern life, from the furniture we sit on to our reproductive choices, to the pills we take, to the foods we eat.

In addition to my family history, I, like so many women, have other risk factors for the disease, including delayed childbirth, a small number of pregnancies, and, because of those two things, many decades of uninterrupted, free-circulating estrogen. I took birth control pills before I was out of my teens. Like most Americans, I have slightly low vitamin D levels, another hazard chalked up to modern life. All told, I'm pretty average, and so are my breasts. In writing and researching this book, I sometimes used my body as a proxy for modern women, testing it for commonly known and suspected carcinogens and holding my breasts up to various scanners, screens, and probes. My daughter, Annabel, gamely signed on for some experiments as well.

At its heart, *Breasts* is an environmental history of a body part. It is the story of how our breasts went from being honed by the environment to being harmed by it. It is part biology, part anthropology, and part medical journalism. The book's publication marks the fiftieth anniversary of two significant milestones in the natural history of breasts whose themes will recur here: the publication of Rachel Carson's *Silent Spring* (which recounted how industrial chemicals were altering biological systems) and the first silicone implant surgery in Houston, Texas, in a woman who really just wanted an ear tuck.

Why should we seek to know the breast better? Why should we care? There are several reasons. One, as individuals and as a culture, we love them and we owe them as much. Two, we want to protect and safeguard them, and to do that, we need to understand how they work and how they malfunction. Three, they are more important than we realize. Breasts are bellwethers for the changing health of people. If we're becoming more infertile, producing increasingly contaminated breast milk, reaching puberty earlier and menopause later, can we fulfill our potential as a species? Are our breasts now the leading soft edge of our devolution? If so, can we restore them to their prelapsarian glory without compromising our modern selves? Breasts carry the burden of the mistakes we have made in our stewardship of the planet, and they alert us to them if we know how to look.

If to have breasts is to be human, then to save them is to save ourselves.

1
FOR WHOM THE BELLS TOLL

A 41-inch bust and a lot of perseverance will get you more than a cup of coffee — a lot more.

— JAYNE MANSFIELD

[Breasts] are a body part that we didn't start out with . . . whole new organs, two of them, tricky to hide or eradicate, attached for all the world to see . . . twin messengers announcing our lack of control, announcing that Nature has plans for us about which we were not consulted.

— FRANCINE PROSE,
Master Breasts

If there's one thing starlets like Jayne Mansfield and Mae West understood, it was the power of their ample endowments. In her 1959 memoir, *Goodness Had Nothing to Do with It,* West writes that beginning in her teens, she regularly rubbed cocoa butter on

29

her breasts, then spritzed them with cold water. "This treatment made them smooth and firm, and developed muscle tone which kept them right up where they were supposed to be." West has good company in doling out ridiculous breast-enhancing tips. On the Internet, you can find creams, pills, pumps, pectoral exercises, even a YouTube video on how to master the boob-inflating "liquefy tool" in Photoshop.

In our culture at least, big breasts get a lot of attention. So I'm told. I display, or rather, don't display, the traditional average American size, a B cup.* Women I know tell me that having large breasts is like walking around with a neon sign hanging around their necks. Men, women, small children, everyone stares. The eyes linger. Some men pant. It's not surprising that some anthropologists have called breasts "a signal." Breasts, they say, must be telling us something about how fit and mature and healthy and maternal their owner is. Why else have them?

* Industry sources in both the United States and Europe contend the average size has increased to a C over the past decade. Mine have creeped up as well. In both cases, I'm afraid, the boost is largely attributed to weight gain.

All mammals have mammary glands, but no other mammal has "breasts" the way we do, with our pleasant orbs sprouting out of puberty and sticking around regardless of our reproductive status. Our breasts are more than just mammary glands; they include a meaty constellation of fat and connective tissue called stroma. To be functional for nursing an infant, a mammary gland need fill only half an eggshell. Big breasts are not required. Along with bipedalism, speech, and furless skin, breasts in their soft stroma-filled glory are one of humanity's defining characteristics. But unlike bipedalism and furlessness, breasts are found in only one sex (at least most of the time). And those kinds of traits, as Darwin pointed out, often evolved as sexual signals to potential mates.

But signals of what, exactly? And does this explain how and why humans won the boob lottery? Many scientists seem to think so, and they have devoted large chunks of their careers to answering these questions. One thing is clear: it's rather fun trying to find out. It's not especially hard to design studies showing that men like breasts. What's much trickier is proving that it actually means anything in evolutionary terms.

I was hoping the answers might lie with

the creative experiments of Alan and Barnaby Dixson, a father-son team of institutionally supported breast watchers. Both based in Wellington, New Zealand, together they've published papers on male preferences for size, shape, and areola color and on female physique and sexual attractiveness in places such as Samoa, Papua New Guinea, Cameroon, and China. Alan, a distinguished primatologist and former science director of the San Diego Zoo, brings a specialty in primate sexuality to their shared project, while Barnaby, a newly minted Ph.D. in cultural anthropology, has a knack for computer graphics and a fresh zeal for fieldwork.

I first met Barnaby on a blustery fall day in Wellington. At twenty-six, his curly auburn hair falling around the collar of a fisherman's sweater, he was very earnest. He walked around with a distracted air and wrinkled brow, and often misplaced things, such as parking receipts. It's not easy being a sex-signaling expert. "Sometimes people think I'm using the government's money to look at breasts. They misunderstand what we do," said Barnaby, who's tall and gangly and speaks with a crisp British accent. As Barnaby pointed out, in places like Samoa, which is now very missionized, it can be a delicate matter asking men to describe which types of breasts they prefer. He said some men think he's "a perv" and get very angry. He avoids men who have been drinking. And in the academic world, grant money can be hard to come by when there are things like breast cancer research to fund. "I probably should have been a doctor," he said. "But I'm quite squeamish really."

Barnaby's latest digital experiment employed an EyeLink 1000 eye-tracking machine and a suite of specialized software. The sixty-thousand-dollar piece of equipment lives in a small, nondescript room labeled "Perception/Attention Lab" in the psychology department at Victoria University. It looked like something you'd find in an optometrist's exam room. You place your chin in the chin cradle and your forehead against the forehead rest. Then you look through little lenses. Instead of seeing an alphabet pyramid, though, your eyes meet images of naked women flashing on a computer screen. If all eye exams worked like this, men would surely get their vision screened in a timely fashion.

On the day I visited the lab, an ecology graduate student named Roan was game to volunteer. Wearing jeans and a baggy T-shirt, he patiently looked through the eye-tracker as Barnaby calibrated the machine. Then Barnaby explained the test. Roan would be looking at six images, all of the same comely model, but digitally "morphed" to look different. Roan would have five seconds to view each image, and then he'd be asked to rank it on a scale of one to six, from least attractive to most, using a keyboard. The images would have smaller and larger breasts and various waist-to-hip ratios. These two metrics, the breasts and the so-called WHR (essentially a measure of curviness), are the lingua franca of "attractiveness studies," which is, believe it or not, a recognized subspecialty of anthropology, sociobiology, and neuropsychology. The theory is that how males and females size each other up can tell us something about how we evolved and who we are.

The eye-tracker doesn't lie. It would show exactly *where* on the body Roan was looking while making up his mind. As Barnaby had explained to me earlier, the machine would measure the travel of Roan's pupils within one-hundredth of a degree, and would record how long his gaze lingered on each body part. "The beauty of the eye-tracking machine is that it allows you to get some measure of behavioral response. You can measure, literally, the behavior of the eye during attractiveness judgment," Barnaby had said.

Roan began staring and ranking. The whole thing took a couple of minutes. He looked a little flushed when it was over.

He stuck around to see how he did. Barnaby called up some neat graphics and computations. A series of green rings overlay the model; they represented each time Roan's eyes lingered for a moment. Some of the rings were on her face, a few on her hips, a whole bunch on her breasts. Barnaby explained them as he reviewed the data. "He starts at the breasts, then looks at the face, then breasts, then pubic region, midriff, face, breasts, face, breasts. Each time the eye rests longer on the breasts." Roan spent more time gazing at the breasts than elsewhere during each "fixation." He rated the slimmer images with large breasts as most attractive.

In other words, Roan behaved just as most men do, and just as Jayne Mansfield knew he would. She could have saved Victoria University a chunk of change.

Barnaby's eye-tracker results may be obvious, but to a scientist, data are key. Barnaby was preparing to publish his study in a journal called *Human Nature.* He believed the work backs up a relatively well-accepted hypothesis that breasts evolved as signals to provide key information to potential mates. That's why men's eyes zoom in on breasts within, oh, about two hundred milliseconds of viewing an image. That's *milliseconds.* "The overall theory is that youth and fertility are important traits when men and women in ancestral times were selecting a partner," said Barnaby. "So it makes sense they'll select for traits that signal mate value, youth, health, fertility." He believes men find breasts useful. Because men liked these informative, novel, gently pendulous orbs — which originally sprang up in the accidental way all new traits do — they selected mates accordingly. The breasty women were the ones who mated most, or mated with the best males, and so the trait was passed down for all to behold. In the world according to Barnaby, that's pretty much the end of the story.

I wondered whether Roan subconsciously sensed that cache of health and youth information in a few seconds of ogling.

"Do you tend to be a breast guy?" I asked him.

"Good question." Roan is a South African who spends his academic time studying rhinos. "Yeah, but not majorly so. It's not like I'm obsessed, like some guys I've met who tend to go on about it. But yeah, I definitely don't have any problem with them."

I couldn't help feeling peeved by the real-world relevance of the eye-tracking study. A man looks at a woman's hips and breasts for five seconds and decides whether or not to mate with her? Was that how it worked in our deep evolutionary past? Was it how it worked now? And even if it were, did it really explain *why* we have breasts in the first place?

"When you're meeting a woman, you're hopefully looking at more than just her breasts," I said to Barnaby and Roan.

Roan blushed and laughed. "Of course! Cheers!"

"That's an important point," interjected Barnaby. "You're not just going to stare at her breasts."

"Some people do," said Roan.

Barnaby felt a need to rescue the conversation. "This is an artificial experiment. It measures what you might call a first-pass filter, just things that are immediately apparent. Then later, when you're meeting and talking, so many other things factor in, like personality, religious background, socioeconomic status."

"Sense of humor?" I asked.

"Yeah," said Roan. "Of course, of course."

Afterward, we ate lunch at the small, creek-side home in the Wellington hills that Barnaby shared with his girlfriend, Monica, a Canadian graduate student studying bird behavior. She made a fantastic soup out of a roasted New Zealand tuber called kumara. A sign above the kitchen read, "Please do not feed the bear."

It turns out I wasn't the only woman whom Barnaby's work made a little uncomfortable and self-conscious.

"Whenever Barny gives seminars on waist-to-hip ratios, all the women run home afterwards and measure themselves," said Monica. (Barnaby's studies and many others have established that men prefer a Marilyn Monroe–esque WHR of .7, meaning the waist is 70 percent the circumference of the hips. Some scientists hypothesize that this magic number represents an optimal level of health and hormones, but the significance of the WHR is highly controversial in the field.)

Barnaby looked mortified. "Yeah, well that's unfortunate."

"I measured mine," offered Monica.

"How did it turn out?" I asked.

"I'm a .75."

Barnaby himself doesn't seem immune to his research. He wears, for example, a beard. In his cross-cultural anthropological studies, he has found facial hair to symbolize masculinity and authority. (His father, who teaches at the university and lives one town over with his wife, Amanda, and an eighty-pound English bulldog named Huxley, sports a bushy white mustache.)

Barnaby's walls boasted several original Alan Dixson drawings, including one of a mandrill and one of a gorilla. Alan illustrates most of his own textbooks, while Barnaby supplies the computer graphics. Alan's latest book is called *Sexual Selection and the Origins of Human Mating Systems*. In addition to their eight coauthored papers, they share a love of animals and a polite, diffident demeanor.

"Barnaby is like a mini-me of Alan," said Monica, laughing. Born in England, the younger Dixson grew up in places like Scotland and West Africa, depending on his father's posts. In Gabon, where Alan ran a primate center and studied sperm competition, Barnaby's family had a pet monkey, a potto named Percy. Living closely among other animals made their behaviors, sexual and otherwise, seem perfectly normal.

Barnaby's older brother is also a scientist. His specialty is an enormous flightless cricket.

Both Alan and Barnaby believe studying mating behavior and sexual selection in primates can tell us much about our own reproductive organs. For example, men have relatively small testicles compared to other existing primates. Alan has written that this might indicate our early human ancestors were polygamous. (On this topic scholars vehemently disagree with each other. The field of evolutionary studies is a blood sport.)

To the Dixsons, enlarged human breasts, like giant testicles in chimps or the orangutan's beard, are "courtship devices" born out of competition and selection. Large testicles produced more sperm, maximizing an individual's chance that his genes, and not his rival's, would penetrate the egg of a promiscuous female. The males with the biggest testicles had more descendants, who in turn had bigger testicles. The Dixsons believe beards and enlarged breasts, on the other hand, are seductive "adornments" advertising genetic quality. Those who attracted the best mates had fitter offspring and, ultimately, larger numbers of descendants, and so the traits persisted. This is the essence of sexual selection as posited by Charles Darwin.

"A lot has been written about what breasts might be telling a guy," said Barnaby. "At its simplest, they're telling the guy that this is a sexually mature woman. Beyond that, there are a lot of hypotheses. One that I find interesting, based on work on Hadza hunter-gatherers in Tanzania, is that there could be a profound preference among men for a nubile breast shape." He explained that as women age and have more successive pregnancies (thus reducing her worth to a new mate), her breasts change. "I'm trying to find a nice way of saying it," hedged Barnaby, "but age and gravity take their toll. The shape tends to lose its firmness and droops somewhat. This could be something that's letting a man know about youth and fertility and potential reproductive output." In other words, guys, go pursue someone a little more worthwhile, biologically speaking.

It's a dog-eat-dog world out there.

The nuances continue. Large breasts sag more than small breasts, said Barnaby, so men likely prefer big ones because they are more "informative" of age. Other studies back up Barnaby's hypothesis, some with real-life experiments. A few years ago in

41

Brittany, France, a twenty-year-old actress of "average attractiveness" with relatively small breasts was given an unusual assignment: to sit in a bar while an undercover researcher recorded how many men approached her. Then she inserted enough latex padding into her bra to bump the cup size up to B and went to a neighboring bar. You can guess the third step: repeat with size C. She wore the same clothes in each bar, a pair of jeans and a tight-fitting sweatshirt. She was instructed to watch the dancing on the dance floor, but not to look at men along the edges. This was repeated for twelve nights over a three-week period.

When she wore the A-cup bra, she was asked to dance thirteen times. When she wore the B cup, she was asked nineteen times. And when her breasts grew to a size C? Forty-four dance cards.

In a similar experiment, Miss Elasto-chest tried hitchhiking by the side of the road, also in Brittany, at the height of summer and during the day. In her A-cup incarnation, fifteen men stopped; in her B cup, twenty men; and in her C cup, twenty-four men stopped. When the passing motorists were women, approximately the same number stopped for each cup size. Another study showed that waitresses with larger breasts get bigger tips.

Steven Platek, an evolutionary neuroscientist from Georgia Gwinnett College, showed college men pictures of breasts while he scanned their brains in an MRI machine. Not so surprisingly, he found the breast images triggered the "reward centers" in the volunteers' brains. "Most of the images capture the attention of the male so much so that it will distract his mental and cognitive processes in ways that could be dysfunctional in other capacities," Platek told me. The *Urban Dictionary* refers to this state as *booblivious*.

Okay, so men are distracted by breasts. All of this sounds familiar to us in Western cultures, but there are problems with making sweeping statements about evolution based on studies about male behavior in pubs. For one thing, I am still hung up on the nubility hypothesis, which might as well be called the sag hypothesis. But speaking from personal experience, I can report my breasts actually got bigger and fuller after pregnancy. I really can't say they are *sagging*, not yet anyway. I am well past the age of what anthropologists call "peak reproductive value." Does a man really need breasts to tell him a woman is getting on in years? Aren't there more obvious signs that don't require awkward social glances? And as anyone who's been to a public shower or springtime college campus can tell you, there is an enormous, and I mean enormous, variety of breast sizes and shapes out there. I'm talking 300 to 500 percent differences in volume, and these are in women of roughly the same age. What other body part is so variable, I ask? If breasts were such important communicators, wouldn't they be more on the same page?

Further complicating the picture, there is also great variety in men's tastes. Barnaby conceded that male preferences aren't as universal as he'd hoped. He expected all men to prefer breasts of a similar size — namely, big. But that doesn't always happen. In his earlier data from the eye-tracker, which he published in *Archives of Sexual Behavior,* the same number of men preferred medium breasts to large breasts, and some men were most enthusiastic about small breasts. And these were all straight, white men from New Zealand. Other studies have shown that Azande and Ganda tribesmen prefer long and pendulous breasts, whereas the Manus and Maasai prefer more rounded ones. One study found that Western men prefer curvier women during a recession, perhaps for their suggestion of comfort and ample calories. In his own study, Barnaby found that men simply liked staring at all breasts, regardless of size or how attractive the image was rated.

If breasts serve as such a great signal of a woman's fitness, so should the areola, posits Barnaby. Younger women who have never had children have lighter areolas, so Barnaby expected men to prefer lighter pigmentation when they rated images in another study. He was surprised to learn that many men like darker, post-pregnancy areolar pigment. Similarly, data on preferences for areolar size were all over the map. And while most men seem to like breasts, in many places breasts are merely pedestrian. Not every culture has a Hooters. The nape of the neck is unbearably sexy in Japan. Bootylicious is where it's at in parts of western Africa and South America. When my son was little, he used to mortify me by going around the house singing a Sir Mix-a-Lot song from the *Shrek* soundtrack: "I like big butts and I cannot lie."

Barnaby knows about these inconsistencies, and they cause him some academic heartburn. But while he acknowledged the data are far from conclusive, he still thinks they hold up. "The amount of visual attention and the amount of evidence that men are attracted to breasts would lead you to think something is going on in evolutionary terms with mate choice and breast morphology."

Barnaby is just the latest in a long line of scientists who have been thinking about how the breasts evolved in step with the male gaze for at least half a century, ever since Desmond Morris published his famous and influential book, *The Naked Ape,* in 1967. (Morris, a British zoologist, is also known for choreographing the gestures and grunts of the actors in *Quest for Fire.*) In *The Naked Ape,* he attempted to explain to a popular audience why humans act the way they do. Describing a prehistoric life very much like the suburban dead-zone of the mid-twentieth century, Morris wrote how out of the Pleistocene emerged "Man the Hunter," unique among primates, who came home after a hard day of stalking beasts and needed his hearth-bound woman to show him a stimulating set of knockers. Without them, he'd have little inclination to stick around and provision the family. (Never mind that hunter-gatherer women procured most of the daily food for their families; that research came later and Morris still has not adjusted his breast-origin hypothesis.)

Since Mrs. Mighty Hunter had to be constantly sexy for this scenario to work, she needed a big front-and-center sexual organ different from what all other primates who did not walk upright on two legs had. Those primates signal sexual readiness, their estrus, with swollen buttocks or labia. "Can we," asked Morris, "if we look at the frontal regions of the females of our species, see any structures that might possibly be mimics of the ancient display of hemispherical buttocks and red labia? The answer stands out as clearly as the female bosom itself. The protuberant, hemispherical breasts of the female must surely be copies of the fleshy buttocks, and the sharply defined red lips around the mouth must be copies of the red labia."

I may never again think of lipstick the same way.

Today, *The Naked Ape* reads like an embarrassing manifesto of male dominance, presented at exactly the same time the women's lib movement was heating up. Just as Linnaeus appeared to be pushing a political agenda in naming us *Mammalia* (nudging women to act more maternal during the Enlightenment), perhaps Morris was too. On the other hand, maybe Linnaeus and

Morris and the whole lot of them were really just breast men.

Clearly, many anthropologists love breasts. In textbook illustrations and museum dioramas, they always seem to depict the latest evolutionary "missing links" with breasts, despite zero fossil evidence for this. Ardi? Lucy? Breasts and more breasts. Even Mrs. Bigfoot is often drawn with a comely pair. We all know men like Morris; there are lots and lots of them. But there are also some leg men out there, like my husband, God bless him. In any case, the science of sexual attraction has been marked by fierce debate and bald accusations of cultural bias that continue to this day.

Try telling some feminist anthropologists that breasts exist because of men, and you might get whacked in the head by a rubber *Australopithecus* pelvis. Elaine Morgan, a Welsh writer, wrote an entire, rather delicious book refuting Morris and his ilk, called *The Descent of Woman*. In it, she thoroughly debunked the notion that the needs of the male drove every clever anatomical adaptation in human ancestors, including breasts. "I find the whole yarn pretty incredible," she wrote. "Desmond Morris, pondering on the shape of a woman's breasts, instantly deduces that they evolved because her mate became a Mighty Hunter, and defends this preposterous proposition with the greatest ingenuity. There's something about the Tarzan figure which has them all mesmerized."

Frances Mascia-Lees, a Rutgers University anthropologist, told me she thinks the scholarship over the past fifty years on breasts and attraction has been a colossal waste of time. "When you talk about the old guys, the same arguments are still being made. They will not die under any circumstance. But when it comes to finding a mate and having children, breast size does not matter, even though many advertisers and plastic surgeons might love us all to think so," she said.

She pointed to a number of holes in the breasts-as-sex-signals theory of origin. If big, firm breasts tell a man that a woman is fertile and ready for sex, then why would her breasts be biggest and firmest when she's already pregnant or lactating? Why is there such huge variation in human breast size and shape, and why are so many women with tiny breasts spectacularly successful at nursing, childbirth, and child-rearing?

Although I hate to admit it, I couldn't help wondering if Mascia-Lees herself has tiny breasts and if that had influenced her contrarian worldview. So I asked her, and it turns out she has the opposite problem. She's a 36DD. When she entered graduate school in 1981, her department consisted of fifteen men and one woman. The American obsession with breasts was annoyingly evident. "Having big breasts meant you were highly sexualized by men," she said. "It was a prickly issue for me trying to be taken seriously as an intellectual." At the time, the Mighty Hunter theory was everywhere. He drove the evolution of the bigger brain, speech, social behavior, bipedalism, the use of tools, and so on. It rankled. It got her thinking. In a sweet-vengeance counterscenario, Mascia-Lees and others instead argue that it's just as likely the female drove these developments, through lactation and the unique demands of the human infant. Just suppose for a moment, gentlemen of the academy, that breasts evolved because *she* needed them, not because her clubwielding cave man did.

Mascia-Lees argues that breasts evolved through natural selection, not sexual selection. It seems perfectly reasonable, if not *more* reasonable, to suppose there was something about having breasts that increased the fitness of the woman and her offspring in what Darwin plaintively calls the "struggle for existence." Male interest, if it even exists universally, was secondary. She posits that breasts helped increase a woman's fat reserves, even if just by a few percentage points. In the poor or unpredictable environment of our early evolution (such as the open plains with their greatly fluctuating temperatures), those extra fat depots could have made the difference in being able to sustain pregnancy and lactation. Humans need to store more fat than other primates because they don't have fur to keep them warm. On top of that, pregnant humans need to mobilize more fat to keep pace with their pudgy babies, whose big brains need specialized stores of long-chain fatty acids. Consequently, women's bodies are designed in such a way that they don't even ovulate unless a body-fat threshold has been crossed. On average, reproductive-age women store twice the fat that men do.

But why store fat in the breasts and not, say, the elbow? Mascia-Lees has a good explanation for this. Fat and cholesterol make estrogen. Mammary glands are filled with estrogen-sensitive cells. We have more estrogen than other primates simply because we're relatively fatter. Here's the sequence: we needed to be fatter at puberty and

51

beyond to produce human infants; our fat made estrogen, and estrogen made our breasts grow because the tissues there are so attuned to it.

In Mascia-Lees's account, breasts are merely "by-products of fat deposition." She admitted her theory is not nearly as testable, or as sexy, as that of the Morris crowd. But that's the point. "I've tried to show that my assumptions are more firmly grounded," she said, "and not just the same cultural assumptions we have now projected back into evolutionary history."

Maybe because I've never had the sort of chest that men stare at, I'm more willing to consider alternative theories of origin. And there are lots. One thing making it tricky is that, unlike the opposable thumb, breasts leave no fossil record. There's no way of knowing exactly when the well-endowed rack appeared in human evolution. Was it before bipedalism or after? Before we lost our fur? Pretty much all of the theories accounting for breasts, Mascia-Lees's and the Dixsons' included, are best categorized as SWAG, scientific wild-ass guesses.

Since breasts are catchments of our collective and individual fantasies, it makes sense that not even scientists are immune from their charms. When we consider the mysterious origin of this fine fleshy organ, breasts become easy metaphors for whatever we desire, from buttocks to political hegemony. One desert zoologist sees in breasts the camel's hump, an adaptation that allows us to survive in arid climates through fluid and fat storage. To feminists, the breast story is a parable of self-determination.

There are plenty of other entertaining, if far-fetched breast-origin stories. Wrote Henri de Mondeville, the surgeon to King Philippe le Bel of France in the early fourteenth century, "The reasons why the breasts of women are on the chest, whereas other animals more often have them elsewhere, are of three kinds. First, the chest is a noble notable and chaste place and thus they can be decently shown. Secondly, warmed by the heart, they return their warmth to it so that this organ strengthens itself. The third reason applies only to big breasts which, by covering the chest, warm, cover, and strengthen the stomach."

In 1840, one physician speculated that fatty breasts warm the milk and "enable women of the lower class to bear the very severe blows which they often receive in their drunken pugilistic contests." He'd perhaps been reading a few too many Gothic novels.

More recently, an Israeli researcher posited that fatty breasts are needed to help the upright female maintain her balance. Otherwise, her fatty bottom would tip her backward. My sister-in-law says this is certainly the reason in her case.

Elaine Morgan, the Welsh critic, has buttressed her own breast theories with some astute anatomical observations. She notes that when our ancestors lost their fur, babies faced some new challenges. Other tiny primates cling to their mother's fur from a very early age. Mom is free to swing from the trees and dig up ants, even while junior breast-feeds. No such luck for humans. We have to hold our little urchins, and the best place for that is the crook of our arm. Even then, though, the nipple still needs to come down a bit to baby. The pendulous breast came to the rescue. Then, once the human baby's hands were free from clutching, they could gesture. An important form of expression evolved and helped make us who we are.

The whole enterprise is greatly assisted by a flexible, unmoored nipple. As Morgan puts it, the brilliantly shaped human breast "ensures that the nipple is no longer anchored tightly to the ribs, as they are in monkeys. The skin of the breast around the nipple becomes more loosely fitting to make it more manoeuvrable, leaving space beneath the looser skin to be occupied by glandular tissue and fat. Adult males find the resulting species-specific contours sexually stimulating, but the instigator and first beneficiary of the change was the baby."

I can wholly affirm that it would be very awkward to breast-feed without a nice moveable feast of a nipple. British anthropologist Gillian Bentley of the University of Durham was nursing her own child when another anatomical light bulb went off: it was our skull shape that drove the ontogeny of rounded breasts. One of the major distinguishing features between us and other primates, indeed between us and most mammals, is our lack of anything resembling a snout. There could be a couple of reasons for this. One is that we have different jaw and teeth structures, the better for eating a varied diet, including cooked meats, which means we don't need huge mandibles to rip apart raw flesh. Another is that we have humongous brains and, at birth, relatively large heads, five times the size of what you'd expect in a primate our size. But in order for newborns to get through our unusually narrow bipedal hips, their faces need to be flat, said Bentley. Flat faces and flat chests don't work well together. Think of kissing a mirror; if the baby's face had to smoosh against a flat chest, it wouldn't be able to breathe through its nose. (Now here you might be clever and ask, as I did, Why didn't evolution instead come up with a different place for the nose, say, near the ear? In fact, why are all mammal noses between the eyes and mouth? The answer has to do with our primitive, born-from-fish infrastructure, a template we're not free to mess with. No doubt it was easier for our genes to tinker with the breast instead.) Thanks to round breasts, we can be smarter.

I started reading more about heads and necks, and I learned about a unique human feature called basicranial flexion. That's how we bend where our neck meets our head, and it is different in us than in anyone else. Human babies, let's not forget, cannot hold their heads up. We may be the only mammal that can't do this. We have unusually big heads, and we also have necks, the better for growing a laryngeal cavity *so that we can speak.* A newborn must be held in order to breast-feed (because we have no fur for him to grab), and his head must be supported, or else his delicate larynx tube, also called a neck, would break. All the more reason why it might be helpful to have a nipple that can come down to the baby. It's a theory, but I like it: thanks to pendulous breasts, we can speak.

Other primates also have fleshy breasts *while nursing,* but without the permanent fat pad they're not quite as enlarged or as round. What's appealing about these woman-centered theories for the breast is that they make some attempt to understand how the organ actually works. The boobs-for-men theories do not.

This is what flummoxes Dan Sellen. He's an anthropologist specializing in nutrition and ecology at the University of Toronto. "Most anthropologists don't study the breast. They have no idea what it does," he told me. "There's a whole industry of folks looking at mate choice, and sure, breasts attract males, but that's different from saying their primary function is to attract mates." Furthermore, he says, "it seems really odd that of all the mammals who have mammary glands, we'd be the only one where the appendage is sexually selected. That would be adding a new function to the breast that's absent from every other mammal."

One could argue that it doesn't really matter why we have breasts. We have them, we love them, they can be useful. "They're pretty, they're flamboyant, they're irresistible," wrote Natalie Angier in *Woman: An Intimate Geography.* "But they are arbitrary, and they signify much less than we think."

But it does matter because, as we've seen, the origin stories wag long political, sexual, and social tails. Beliefs about the origins — and thus "purpose" — of breasts can even influence their health and functioning. It's not just the feminists who are down on the sexual selection stories. Sellen is also, because, as he put it, oversexualizing the breast detracts from infant health and contributes to body image problems in young women. It's hard enough to get women to breast-feed as it is. "If we keep reinforcing that breasts are exclusively for sex, we're always undermining the idea that breast-feeding is normative and normal and should be supported. Look," he said, "the reason humans have a slightly different breast structure has to do with delivering essential nutrients."

As a specialist in infant nutrition, Sellen acknowledges his own biases. But the ones governing the work of the Dixsons and the other Morris descendants are stronger, he argues; in fact, they're rooted in human

58

nature. "Humans will make anything sexy. We can transfer some kind of sexiness to any trait."

"Like bound feet?" I asked.

"Exactly," he said. "In cultures that start to hide women's bodies, that can explain why men are attracted to these traits. With breasts, men are just loading culturally a set of symbolizations onto something that really evolved for more direct reasons. We've got to be more scientific about it."

These academic fisticuffs were very much on my mind in Wellington. On my last morning in town, I joined Barnaby and Alan Dixson for coffee. Alan was wearing a pink button-down shirt, suspenders, and a tan blazer. A Maori fishhook made from cow bone hung from a cord around his neck. He was part gracious Englishman, part eccentric Englishman. With his bushy mustache and slightly wild white hair, he reminded me of some of the primates he has spent so many years studying.

I asked Alan if he thinks it is possible that natural selection, not sexual selection, was driving the evolution of breasts. "I think the two went lockstep," he answered judiciously. "Laying down the fat is naturally selective, because you need it. Then it becomes a question of where to put it. If you're a dormouse, you put it in your tail. If you're a mandrill, you put it in your ass." Now the coffee was kicking in, and engaging professor mode was fully launched. "If you're a human, and you put it in your chest, then maybe it's sexually selected too, because in younger women, you'd have the appearance of healthy physiognomy. Men might prefer women with these attributes. We're talking about a dynamic process. We're not talking about the peacock's tail, which is no bloody use at all. We're talking about something that displays underlying health and well-being. I imagine there's something to do with more than just lactation and pregnancy. I imagine breasts have something to do with displaying readiness for reproduction."

"Yes," mused Barnaby, hunkering over his espresso and wrinkling his brow again. "These things may be linked together. It's a fair point."

Ah. It seems we'd arrived at a happy medium. I could go home now. Except, for some reason, I still found myself not altogether satisfied. The more I thought about it, the less it seemed that sexual and natural selection of the breasts arrived in lockstep. In fact, I became increasingly convinced that breasts have been categorically miscast in modern history.

I kept thinking as my plane thrummed out over the Pacific, which long ago men and women crossed in rickety dugouts following their human dreams of migration and survival. Then, as for all of our history on this planet, they fell in love or in lust, and everyone who could have children did.

What if instead of men selecting breasts, the breasts selected the men? It's possible that once upon a time, Early Man loved lots of different specimens of Early Woman, some with no breasts, some with small breasts, some with hairy breasts, whatever. Man, as we all know, is sometimes not that picky. Then, for the reasons described earlier — fat deposition, cranium shape, the development of speech, and the long neck — the women with the enlarged breasts and their infants gradually outlasted the others. That is, after all, the way natural selection works.

Consequently, the people who could talk and sing and have the biggest, best-fed brains were the ones born of women with breasts. It makes perfect sense that we would grow up to appreciate and enjoy breasts, eventually putting pictures of them in eye-tracker machines in universities.

Perhaps, all along, the breasts were calling the shots.

2
CIRCULAR BEGINNINGS

. . . from so simple a beginning endless forms most beautiful and most wonderful have been, and are being evolved.

— CHARLES DARWIN,
On the Origin of Species

Hopefully now we can all relax and understand that breasts are, truly, designed for the purpose of feeding infants. With that out of the way, it's worth exploring briefly how the evolutionary zinger of lactation came about.

As wondrously unique as human breasts are in their pendulosity, their basic glandular architecture is shared by all other mammals. Our packaging is just more fetching. Other mammals have some notably oddball features. The manatee has nipples under her flippers. The nipples of the aye-aye (a small primate) sit near the mother's rear end. The gelada monkey's nipples are so close together that the baby can suck both at once. Spiny anteaters and platypuses, rare egg-laying mammals, have no nipples, but they "sweat" milk to the puggle in their pouches through special glands. I don't know what a puggle is, but I want one. A hedgehog-like mammal from Madagascar takes the trophy for most nipples, twenty-four, and the Virginia opossum is unique in having an odd number, thirteen. The only male animal believed to lactate is the Dayak fruit bat, but even that is somewhat contentious — it's unknown if the substance has any nutritional value.

But we're more alike than not. Evolutionary biologists point out that the six thousand or so genes governing lactation are among the most strongly conserved ones we have, meaning they haven't evolved as recently as genes governing, say, hair or toes or the ability to digest Cherry Garcia ice cream. We have protected our primitive lactation genes because they have served us so well. If the ability to lactate is among our most valuable genetic assets, the fat globule is its crown jewel. At its core, lactation is a fat-delivery system, very little changed over millions of years, except for some dietary tweaking. Each mammal has its own proprietary ratio of fats to carbohydrates to proteins. Human milk, for example, contains one-sixth the protein found in that of the quokka (a small marsupial), and one-fiftieth the fat found in seal milk.

All mammary glands, ours included, serve four basic eon-tested functions: First and most obvious, they provide specialized, highly adapted chow for each little newborn mammal. Second, they provide immune support for same. Third, getting a little more subtle, they produce hormones that work as natural contraception, ensuring that a mother's births are spaced adequately apart. Finally, they provide a "window of learning" in which young mammals can focus on acquiring skills rather than desperately seeking breakfast. (Note: attracting the opposite sex was never part of the original job description.)

If it works, don't fix it. And for millions of years — even hundreds of millions of years — it worked unbelievably well. In fact, mammals' ability to lactate was crucial to our success. It was an innovation that changed the world.

The earliest lactating species showed up at the end of the Triassic, about 220 million years ago. Before that, animals popped out of eggs and then immediately had to find food. It was rough out there. Dinosaurs ruled the earth. By the start of the Cretaceous, 135 million years ago, dinosaurs and giant sea monsters still held dominion, and the few mammals scampering around were small and rodent-like. But around 60 million years ago, at the beginning of our Cenozoic era, some very dramatic things happened to the earth's temperature and moisture levels. Maybe a meteor hit the earth; maybe a volcano erupted; maybe the climate simply hiccupped of its own accord. Whatever it was, nearly 40 percent of all creatures went extinct in pretty short order. Dinosaurs? Kaput. Sharks and large marine reptiles? For many, ditto.

The new world order was cuter and furrier and made up of strong social bonds, a keen sense of smell, and a lot of snuggle time. Mammals owned the Cenozoic.

The dramatic emergence of lactating species has been something of a thorn in the side of Darwinians, who had a hard time explaining it. Critics of evolution like to point to lactation, along with the development of eyes, as events that couldn't simply evolve gradually, by accidental mutation, in a way that conferred an immediate survival advantage. How could you have a partial eye or a partial teat? Darwin himself went out on a limb to speculate that mammary glands slowly evolved from sweat glands in brood pouches where some fish and other marine animals kept their eggs. The sweat gave the eggs a little extra nourishment, and the system was off and running.

It turns out that Darwin hit pretty close to the target. At least so says Dr. Olav Oftedal, the closest thing the planet has to an expert on the evolution of lactation. Oftedal came to his specialty in a circuitous way. The child of a Norwegian diplomat, he grew up chasing snakes and playing in the hardwood forests of Europe and the United States. But in the late 1960s, wanting to be socially relevant, to change the world and improve infant nutrition, he started working for aid programs in the developing world. Oftedal grew discouraged with the pace of policy work, though, and went back to school to study maternal-offspring nutrition, this time in the animal kingdom.

"It was amazing how Darwin hit on things, and this was before genetics!" boomed Oftedal from his office at the Smithsonian Environment Research Center near Chesapeake Bay in Edgewater, Maryland. Oftedal has built his three-decade career studying the remarkable lactation habits of seals, bears, bats, and monkeys, among others. Weddell seal pups near Antarctica must *qua-*

druple their weight in the first six weeks of life, so seal milk is around 50 percent fat, among the fattiest known. A stiff wind might turn it into butter. Lactating seals have unique feeding pressures; in some species, the mother nurses for a few days, then takes off for days or even weeks to replenish her body stores far from home. The pup has to wait for her return.

Oftedal has tasted this wondrous seal milk, declaring it "fishy." You might think obtaining a sample is no simple matter, and you'd be correct. He throws a rubbery bag over the mother's head, and then pumps her teats with a handheld device — all in temperatures around 20 degrees below zero Fahrenheit.

Oftedal sees lactation as a quintessential competition between mothers and offspring for nutrition. In the case of the Weddell seal, the mother is nearly depleted by the needs of her fast-growing pup. Seen in this light of tremendous costs, lactation wouldn't have evolved if it were not very, very useful. It also necessitated radical innovations in both the mother's hardware and that of her young, including different teeth and brains. It was not to be embarked upon lightly.

First, we had to figure how to make the hardware, the mammary gland itself. Teeth, oddly enough, offered a blueprint. Having developed much earlier, they pioneered a technique for simple bioorigami, showing how two layers of tissue could fold in on themselves and make proteins to build an organ. In a sequence that would make dentists everywhere happy if they knew it, we would never have breasts if we didn't have teeth. But it was still a long way to get from a molar to a milk machine.

For one thing, we had to keep upgrading the software. Metabolic activity is regulated by hormones that flow back and forth from the brain to target cells all over the body, including the mammary gland. As mammals evolved, so did the complex hormonal conversation necessary to regulate their changing bodies. Mammary glands evolved receptors on their cells to "listen" for and collect estrogen, progesterone, prolactin, lactogen, and many other hormones. These tell the glands when to mature and when to regress. They reveal when there is a fetus in the oven, when to deploy a glandular growth spurt, when to shut down milk production, even what sex the fetus is in order to fine-tune the composition of the milk.

"We don't realize how strange mammary glands are," said Oftedal. "There are tons of placenta-type structures out there in sharks and lizards, but there's nothing else like mammary glands. To have a skin gland producing a large flow of liquid rich in nutrients is very strange."

Darwin did not have the advantage of being able to date fossils, so he couldn't know how far back lactation went. It's very far indeed. Lactation even predates mammals, which now sounds bizarre since it is the defining characteristic of them. Here's how Oftedal thinks it happened. Once upon a time, long before there were mammals, there were mammal-like reptiles called synapsids. These split off from other reptiles and proto-dinosaurs around 310 million years ago. Instead of scaly skin, synapsids had leathery, glandular skin containing hair follicles. They developed specialized teeth and a unique jaw that would someday evolve into a mammalian palate, nose, and ear bones. Early synapsids looked like giant terrestrial lizards. After thriving for tens of millions of years, the synapsid-descended therapsids were nearly wiped out — along with 70 percent of all creatures — by the Permian-Triassic mass extinction event 250 million years ago. Fortunately for us, a few survived, evolving into small, mammal-like creatures called cynodonts, but they would soon be eclipsed by emerging dinosaurs in the late Triassic and the Jurassic period. Somewhere in here, lactation proper began.

The proto-mammals had kangaroo-like pouches that transported eggs, then hatchlings. The eggs, said Oftedal, were made of leathery shells with porous, parchment-like coverings. Because they were porous, they lost moisture easily and were susceptible to harmful microbes. But the mother could help solve these glitches. Maternal skin glands in the pouch began to secrete fluids and fight germs. The first fluid was a sort of natural Lysol. It wasn't much of a leap for nutrients to eventually find their way into the mix. Eventually the happy hatchlings had constant, enriched fast food: Lysol on a burger. It's worth mentioning again that mammary glands likely first evolved for immune support (stay tuned for more on this in chapter 9).

Out of the hostile, climate-addled Cretaceous period, only a few small, straggler mammals, about eighteen genera in all, survived. They adjusted their skeletons and airways for better running, and they became nocturnal. They also invested more time and energy in their young and internally regulated their temperatures in ways the reptiles could not. Lactation enabled most of these changes. Endothermy, or body-temperature regulation, for example, wouldn't be possible in tiny offspring without the fast metabolism offered by specialized, high-fat milk and intensive, snuggling parental care.

"Little creatures that had become warm-blooded and active and lactating when the world went to hell were in a better position to survive," Oftedal explained. "The consequence of lactation was that ultimately you could defer becoming an adult, when you have to kill or find your own food." Consequently, mammals could become much more specialized because they weren't forced to stay in a habitat that provided kid-friendly food. They could transform adult food into milk. Ruminants, for example, evolved eating stuff that their babies could never handle. Another example Oftedal offered are whales, which spend part of the year getting fat by feeding in the rich polar regions, then migrate to the warm but food-scarce tropics for birthing and nursing. "They do this because they can lactate!" he said, getting more excited now that we were talking about marine mammals. Crocodiles, by contrast, are stuck by the riverbank all day long so the baby crocs can go fishing. Another benefit of lactation is that mammalian babies' heads can start smaller (because they don't need teeth) and later grow larger to accommodate more specialized teeth and bigger brains. Being born with a small head and body is also very helpful for the mother's mobility.

The key concept here is flexibility across habitats and niches. By the late Paleocene, there were a hundred genera of mammals made up of thousands of species, from saber-toothed tigers to hornless rhinos to flying bats to primates. The largest known mammal, a rhino-like *Indricotherium transouralicum,* walked Eurasia 34 million years ago. It weighed forty thousand pounds. The bumblebee bat, on the other hand, stands one and a half inches tall. As Oftedal put it, "You can have the tremendous diversity of form of things like cheetahs, buffalo, mice, manatees, seals, all made possible because of lactation!"

The bonuses continue. Because the young must stay with the mother all day, "you have cultural transmission!" The offspring learns from the mother. Because mother and baby must communicate and "love" in some form, the mammalian brain's six-layer neocortex evolved (together with a new sensitivity to hormones), making possible an acute sense of touch, sound, and smell, and eventually conscious thought, reasoning, and language.

Lactation, with its tremendous metabolic efficiencies, made possible the huge difference in brain volume — up to ten times — between reptiles and mammals. The need for suckling drove the development of the palate and tongue muscles. These developments in turn prepared the way for the evolution of speech in certain higher-order primates, namely, humans. Lactation enabled complex communication.

"You have the evolution of highly social behavior!"

Oftedal spoke with the enthusiasm of a convert, but, surprisingly, there is barely a mention of lactation in many textbooks about the evolution of synapsids and proto-mammals. No one pays adequate attention to lactation, said a wistful Oftedal, even though it's perhaps the single most earth-shaking event in mammalian ascendance. "It's because the field is dominated by men who don't think much of breasts except as sexual objects," he laughed.

Not that again.

But he's right. The evolutionary biologists have spent so much time on the unusual and mesmerizing appearance of breasts that they've forgotten there's something profound and fundamental inside them. Students of the breast haven't much noticed that these mysterious interiors evolved to be intricately connected to the rest of the body and to the outer environment. Mammals could not have adapted so well to their changing planet if their glands weren't constantly checking out the neighborhood. Need some Lysol? Here's a squirt! Need a head that's five times bigger than your ancestors? Cheers! To be sure, this adaptation was the result of natural selection, and plenty of mammary glands out there didn't make it. But thanks to all that hard-won evolution, the gland itself is also capable of making minute adjustments on a day-to-day basis to serve mothers and offspring. Our blind spots to the breast's amazing evolution may help explain why there have been, and continue to be, big gaps in our knowledge about how these organs work. Fortunately, some scientists are taking a deeper look.

3
Plumbing: A Primer

I have heard a good anatomist say, "the breast is so complicated that I can make nothing clear of it."

— SIR ASTLEY PASTON COOPER,
On the Anatomy of the Breast

In antiquity, a temple on the Island of Rhodes displayed a goblet said to be molded from the perfect breasts of Helen of Troy. Her face may have launched a thousand ships, but it was her breasts that really buoyed the army. In the Middle Ages, French King Henry II reportedly had casts made of the "apple-like" breasts of his mistress Diane de Poitiers for his wine cups. Marie-Antoinette's breasts were believed to inspire the design of shallow French champagne coupes (not the narrow fluted ones, heavens), as well as of some celebrated porcelain milk bowls made by Sèvres.*

Some people are just born with the blue-prints for a great pair of knockers. It all has to do with a magic ratio of ligaments to fat to glands. Human breast tissue falls into three large categories: fat, stroma (connective tissue, mostly), and glandular tissue, called parenchyma, made up of ductal epithelial cells. On a mammogram, the light parts on the image represent the "gland" part. The dark parts are the fat. In humans, the gland is made up of numerous ducts snaking through the fat and stroma like tendrils of fireworks in the night sky.

If you want to learn how a car is made, you go to the assembly plant. If you want to know how breasts develop, you visit Zena Werb, a cell biologist in the anatomy department at the University of California, San

* The French royalty evidently had a thing for their organs. It is believed that Napoleon's penis was removed from his body for posterity. In the 1920s, it was displayed in a blue velvet case at the Museum of French Art in New York, where one observer described it as looking "like a maltreated strip of buckskin shoelace or shriveled eel." It was offered for auction in 1981. It did not sell, leading one British tabloid to trumpet, "Not tonight, Josephine!"

Francisco.

The first thing she'll tell you is that we don't know much.

Oh, we know some things, like basic breast dimensions. The average breast weighs just over a pound, but this can double in late pregnancy. Its mean volume is about 2/3 cup, or 561 milliliters. (The woman said to have the largest implants in the world, though, wore a size 38KKK, the fluid equivalent of 2.6 gallons, or 21 pounds. She reportedly contracted a staph infection in her breasts and had to get the implants removed. As one news outlet put it, "What goes up must go down.") Over the course of a menstrual cycle, breast volume varies by 13.6 percent, owing to water retention and cell growth. Some studies have found that the left breast tends to be bigger than the right breast. In any case, one breast is usually, on average, 39.7 milliliters, or nearly a fifth of a cup, bigger than the other.

For decades, the standard way to measure volume was to make a plaster cast of each breast, then fill it with sand "of known density." Bra manufacturers prefer to use standard mathematical formulas. They take into account radius and diameter and cone shape versus hemisphere shape. They look like this:

$V = \dfrac{D^3 \times .5236}{2}$	$V = \dfrac{r^3 \times 4.1888}{2}$	$V = \dfrac{4\pi r^3}{2}$

These equations are taken from a mechanical journal published in the wake of Sputnik. Its proud author declared, "Brassiere design is one engineering activity, at least, in which the United States is ahead of the Soviet Union."

Werb is not interested in measurements. She is unlocking the mysteries of how breasts develop. While visiting her at her small office on a high floor of the UCSF Medical Center, I saw breasts everywhere. Werb wore a broche of gold concentric circles pinned to her black jacket. She had on gold hoop earrings and round glasses. A self-portrait of Frida Kahlo lactating hung on the wall in a frightful *Madonna lactans* twist that only Kahlo can do. Against another wall leaned a glow-in-the-dark skeleton. Life and death. Breasts confer both.

Unlike any other organ we have, breasts do most of their developing well after birth. In other complicated organs, such as the brain, the penis, and the testes, the basic architecture is laid down at birth. But the breast has to fully build itself out of nothing during puberty. Even then, it's not done.

The gland grows new milk-making structures under the influence of pregnancy hormones. Once an infant has weaned, a switch flips somewhere and the gland shuts down and shrinks. The breast must construct and then deconstruct itself over and over again with each pregnancy. It's like Caesar's army, making a camp city and then breaking it down on its relentless march across Gaul. Even if a woman never gets pregnant, her breasts pack and unpack a little bit each month just in case.

The developing breast is a challenging organ to study because, as Werb pointed out, you can't easily find adolescent tissue samples to cut open. After a twelve-year-old girl has been killed in a car crash, you can't just go up to the parents and say you want her breasts. "There's a question of delicacy," she said.

So for the most part, Werb does what many of her breast-researching colleagues do. She spends a lot of time with mice and rats. Among other things, she's discovered how glandular tissue grows, literally, molecule by molecule. This fruiting out is called mammapoiesis, a word that makes me think of breasts reciting poetry, which seems somehow apt. Werb has actually caught it on video. It's breast porn of a whole new kind.

She showed me the action footage of molecules becoming glands becoming breasts. Brilliant green dots appeared, forming a kind of shoreline. The dots, tiny rings at the front of a shoreline, slowly bubbled outward. They were the little milk ducts tunneling forth as they grow during puberty. It looked so innocent. But then she described the shoreline as "the invasive front." I felt like I was watching a reenactment of D-day, and in a way, I was. Ductal cells grow into surrounding tissue by massively proliferating. If that sounds familiar, that's because it's essentially the same process that occurs in cancer, except here, it's supposed to happen. At some mysterious signal in late childhood, the milk ducts set out on their journey, forming dense, veinous branches. The surrounding tissue, mostly the stroma, including material called the "extracellular matrix," permit the trespassing ductal cells to cross.

Werb's work illustrates how the different cells of the breast are constantly communicating; the extracellular matrix must allow the gland to grow through it, possibly because the glandular cells command it to. In cancer, a tumor sends out similar signals.

If we can learn more about the molecular stop and go signals of development, we'll have more clues for treating cancer. Essentially, it looks as though cancer cells just think they are making another breast.

Werb's images were too high-tech and abstract for me. I needed to step back from the microscope to see the bigger picture. Thankfully, I had my handy copy of Sir Astley Cooper's *On the Anatomy of the Breast,* originally published in 1840. Back when he was writing, it was easier to scrounge up dead people's organs. As a result, some of the best work on breast anatomy is 170 years old. (This begs a digression. In England, the king granted surgeons free use of the organs and bodies of up to a hundred executed criminals a year. But soon that wasn't enough — and it didn't include many breasts in any case — so medical schools turned to the cadaver trade. It wasn't technically illegal to take someone's dead body, because corpses, unlike the clothes attached to the corpse, weren't considered property. Grave robbers, so-called resurrectionists, became so prominent they actually organized unions and went on strike for better corpse prices. After a huge scandal involving a Scottish boardinghouse whose proprietor murdered drunken tenants for the body trade, Parliament passed the Anatomical Act of 1832. This required anyone performing dissections to have a license and to use only bodies that were donated or left unclaimed in prisons and workhouses.)

Born in 1768, Cooper was a gifted British physician. In 1820, he achieved fame (and a baronetcy) after removing an infected cyst from the head of King George IV. He became surgeon general to the king and later to Queen Victoria. In his sixties, Cooper turned his attention to "senology," or the study of the breast, from *seno,* Italian and Spanish for "bosom" (not to be confused with Sinology, the study of China). However he got them, Cooper harvested breasts from dead females (and quite a few males) of every age and inclination. It sounds morbid (okay, it *is* morbid), but Cooper found these dismembered breasts very beautiful. He became their most faithful and celebrated chronicler and would would lend his name to a set of bosomy ligaments.

It's not surprising that doctors were fascinated by breasts. For one thing, they were a known source of tumors. But most incredibly, these organs performed the transmutation of blood to milk. Even Jesus Christ could work with only loaves and fishes. How did this alchemy work?

Cooper knew it worked slightly differently in different mammals. Cows have one major channel flowing out of the teat. It's the Erie Canal, compared to the Nile Delta in humans, rivulets that emerge from the nipple through tiny holes like those at the tip of an old-fashioned watering can. To study the complex ductal system, Cooper injected more than two hundred (dismembered) breasts with wax dyes or mercury in a scientific quest one modern Scottish anatomist called "a breadth of experience unparalleled before or since." Cooper then drew detailed "galactograms," or images of the milk duct system, which he published in his definitive text. The renderings are available on the Internet (as are so many less scientific images of breasts), and they are fascinating. The ducts appear rubbery and stringy and convoluted — seaweed salad in a bowl. For his reader's edification, Cooper also includes plates showing The Udder of the Ewe and the Dug of the Ass.

The descriptions of Cooper's "preparation" read a bit like an old farmhouse cookbook: "it is requisite that the breast be put for a short time in boiling water, when the skin and fat become detached, and the gland, like other albuminous compositions, is left extremely hardened . . . dried, after being boiled, the gland may be preserved for many years."

Cooper labored under no delusions about who or what breasts were for: "in all the class Mammalia, [Nature] has provided glands to supply bountifully, by the secretion of milk, that nourishment which the young animal will require soon after it begins to breathe. The Breasts, or Mammae, are formed for this purpose." This perspective does not, however, keep him from often referring to the breast's elegance and pleasing appearance.

Through his meticulous dissections, Cooper learned more about the breast than just about anyone before or since. Among his many astute observations was that the blood-to-milk miracle happens in the alveoli cells deep in the gland, in tiny grapelike structures that form lobules. The lobules merge into separate ductal networks called lobes. If the lobules are the grapes, the lobes are the vines. These make up the breast's basic dairy equipment. The number of lobes varies among women, with the average being two dozen. Each lobe empties out into an orifice on the nipple, and sometimes different lobes share one orifice. The average nipple has about twelve orifices. When the gland is developing in childhood and puberty, growth starts at the nipple and branches backward toward the chest wall, and during pregnancy the lobules with their alveoli finally form.

Cooper did not know about the breast making stem cells, but he discovered many other wonders, such as the ability of the nipple to secrete substances other than milk, including protective oils. He saw that the areola has little bumps to help form a seal with infant lips. He recognized the glands are densely wired with veins and nerves, all the better for responding to infants and stimulating lactation (this is also why the breast is so sexually responsive, but Cooper — ever the Victorian — didn't write specifically of its erotic capacity). He examined the properties of breast milk and even made cream cheese from it. He told his medical readers how to distinguish a benign from a malignant tumor, and how and when to operate.

Cooper noted that sometimes men grow breasts, and on rare occasion have been able to produce milk-like fluid. This happened to a twenty-two-year-old soldier he examined. When male or female breasts produce milk unrelated to pregnancy, the condition is called (rather uncharmingly) galactorrhea. He correctly observed that man-boobs are usually made of fat and not gland, but he saw that there are small amounts of glandular tissue behind the male nipple, including, sometimes, ducts. We now understand this to be the result of unusual hormonal influences such as a pituitary disorder or environmental exposures. Newborn babies also sometimes produce milk, called witch's milk, the result of maternal hormones coursing through the baby.

Why do some men have breasts, and moreover — that perennial biology question — why do men have nipples? Cooper understood that both males and females are equipped with the same hardware early in fetal life, but he didn't speculate on how they grew distinct. Here's a newer bit of developmental biology: After an embryo is conceived, it is rigged to become either sex. This is called its bipotential state. During its first six weeks, certain pre-organ structures are laid down, including two parallel milk ridges. Born of ancient genes and common to all mammals, these ridges run up and down the length of the torso. If the fetus inherits female XX genes and the process unfolds in the expected way, then estrogen will turn the primitive plumbing into a female reproductive tract. If the fetus inherits male XY genes, testosterone will inhibit that progression. Since a later dose of estrogen can make male breasts grow, and then a hit of prolactin can fire up milk-making, it would theoretically be possible to have fathers become full partners in lactation. They could take a milking pill. But good luck with that one.

In animals with large litters, the milk lines launch multiple teats on each side. Primates, elephants, horses, cows, and some other mammals get just one set, usually located toward the hind legs. In about one human in a hundred, a vestigial extra nipple or two can show up. Cooper was familiar with these cases, which is why, in his book, he writes, with characteristic judiciousness, "the breasts are generally two in number."

And of course, Cooper knew all about the inconstancy of breasts. They grow from almost nothing, gradually during childhood and then quite rapidly through adolescence, pregnancy, and lactation. The pace of change slows down again through peri-menopause and menopause. The ligaments bearing his name tend to relax over time, and the volume of tissue often decreases as the glandular lobules atrophy (stay tuned for more about this in chapter 13). So, yes, there really is a sag factor, but when and how it happens vary among individuals. The nipples, too, change from small and light in youth to larger and darker in adulthood. From the time we are born, our breasts are on the move.

Cooper did such a thorough job investigating the breast that for the next century or so, no one bothered to learn any more. Until recently, the greatest advances in understanding the mechanics of the mammary gland were made in the dairy field. As to the anatomy of the breast — and the effects on said anatomy from the inexorable march of time — that would eventually undergo some eye-popping revisionism.

Thanks to recent technology, you no longer have to be dead to have someone inject foreign substances into your mammary glands.

4
FILL HER UP

... but on the fourth night, Ormond chancing to praise the fine shape of one of her very dear friends, Miss Darrell whispered, "She owes that fine shape to a finely padded corset."

— MARIA EDGEWORTH,
Ormon

Breasts might exist for the purpose of feeding infants, but let's face it, for most women these days, breasts fulfill that destiny only briefly, if at all. The rest of the time, they sit around trying, sometimes desperately, to look nice. In other primates, "breasts" exist *only* while lactating. For us, lactating is beside the point. Many of us will think nothing of jeopardizing lactation at that other altar of evolution: beauty. Throughout the ages, women have alternately flattened them, buttressed them, veiled them, decorated them, and bared them, sometimes in the course of a day. Now, with enough cash or credit, we can change them for life.

According to the American Society for Aesthetic and Plastic Surgery, 289,000 women went under the knife to enlarge their breasts in 2009, vaulting it to the country's most popular cosmetic surgery ahead of nose jobs, eyelid lifts, and liposuction. That figure does not include 113,000 breast reductions in women, 17,000 breast reductions in men, 87,000 "breast lifts," and 20,000 implant removals. The history of how we got from A to B, or DD, as it were, is a sordid and fascinating tale of marketing, mass hysteria, and environmental disease. To see where we've ended up, as well as where it all began, I went to boob job ground zero: houston.

At the swank office suite of Dr. Michael Ciaravino, over eight hundred pairs of breasts a year get the full Texas treatment: silicone, mostly, and a smattering of saline. Ciaravino is a true scion of a storied boob-job lineage, having trained with the doctor who trained with the inventor of implants. An energetic forty-five-year-old, he performs more augmentations by far than any doctor in Texas. His office is where Trump Plaza meets Jiffy Lube. I walked into the white marble sanctum on a crisp winter day.

Reflecting the notion that boob jobs are as much about consumerism as medicine, tasteful displays of cosmetics, Ciaravino T-shirts, and a giant poster advertising MemoryGel by Mentor for "superlative enhancement" met me just inside the glass doors. Add to this soft lighting, spotless white and taupe furniture, and arresting megaphotos of women in expensive lingerie.

Dr. C, as he's known affectionately by staff and patients, had agreed to walk me through the experience as if I were a regular patient. It was all so real, so slick and seductive, so full of metaphorical lotuses that I almost left with a new titanic rack. First, I was greeted by Katye, who genuinely fits the description of blonde bombshell. Like many of the curvy and silken-haired assistants here, she's been either a swimsuit model or a professional cheerleader. We practically sashayed to a corner office overlooking leafy west Houston, not far from the Galleria mall. Several curvy vases accented the room's modernist décor, suggesting shapelier times ahead.

"Welcome to the practice!" Katye began. She told me that Dr. C has been practicing for fourteen years and "has been able to perfect the technique." She showed me a book of before and after photos, in which (mostly) perfectly nice breasts end up looking like water balloons on a skinny rib cage. These headless torsos did, I have to admit, look much more sexed up in the after shots, since by now we've all been conditioned to associate big fake breasts with sex. More on that later.

Katye walked me next door to the 3-D imaging room, where, in the name of journalism, I disrobed. After I comfortably settled into my white waffle-weave robe, she showed me implant samples. They were about the size of a large Krispy Kreme. Both the silicone and saline ones were cased in a round, clear, silicone bag. The silicone implant felt nice and soft in a detached way, like bread dough through Saran Wrap. The saline one felt like a bag of water, which is what it is. Women who wear these sometimes make sloshing noises, and ripples can show through the skin. They are less expensive, though, and may be safer if the implant ruptures. In that case, the breast deflates like a flat tire. When a silicone implant ruptures, it is supposed to stay in place since it has the viscous properties of a gummy bear. This is a vast improvement over the more syrup-like silicones of old.

Dr. Ciaravino came in and introduced himself. He has a broad, tanned face and shoulder-length brown hair. He wore a white lab coat and a thick neck chain. I could easily see him relishing his pastimes, which, according to the office literature, include driving a Porsche and playing electric guitar. I channeled my inner Houston housewife. I told him I'd borne two children, had breast-fed for years, and after going through life as a size B, was now curious as to what life might be like as a C. He nodded sympathetically. "Let's have a look," he said.

The robe came off, and Ciaravino pulled out a small tape measure. He measured me from collarbone to nipple, from nipple to under-breast fold, and from nipple to nipple, calling out numbers to Katye. He took a step back and mashed my breasts together with his hands, then squeezed each one like a club sandwich. I felt like I was awaiting the word of St. Peter. I was secretly hoping one of the world's foremost experts on flawed breasts would be so vexed by my nice, very normal breasts that he'd tell me he had nothing to offer.

"Well, first off," he began, "let me say you'd be a great candidate for breast augmentation." He assessed me some more. "Where you're lacking a little is some upper fullness here," he said, referring to the slope above my nipples. "You actually have a decent amount of breast tissue to begin with. We just need to give it a little boost. Silicone would really serve you best. What I would say if we were truly just trying to gain a little upper fullness and enhance its look, we would want to work with implants in a 250 to 275 cc range. This would move you into an average C size." (Silicone implants from Mentor, for which Ciaravino is a paid consultant, come in a range of about 100 to 800 ccs, or cubic centimeters. Most women in Texas go much bigger than what he was recommending for me. "Big breasts are part of the Texas tradition," he said. For perspective, some women test sizes by filling sandwich bags with rice: 275 cubic centimeters is the equivalent of 1 1/5 cups of rice; 800 cubic centimeters is almost 3 1/2 cups of rice.)

Ciaravino then led me to his new $40,000 Vectra imaging machine, which would simulate how the implants would look in my breasts. He ducked out, and I, still half-naked, stood motionless in front of the small-saguaro-sized device, with its white plastic trunk and arms, while it captured me in 3-D. Katye clicked a mouse on a computer and then told me I could get dressed behind a small curtain. Soon an image of my torso popped up on her monitor, and together we watched while she punched in some magic codes. Two images appeared on the monitor, me with my real B-plus breasts and then me with big breasts getting bigger and bigger.

"Oh my God," I said to the screen. I was va-va-voom. But not in a good way. My breasts were big and pendulous and pointing outward. My nipples had the strabismic look of a walleye.

Dr. C popped back into the room and looked at the monitor.

"Oh, that's huge," he said.

"I kind of have a sideways thing going on," I said.

"Yeah, that doesn't look too good. I would back it up to about half of that," he told Katye at the controls. "Keep going, keep going." My cyber boobs were shrinking before my eyes. "Sometimes the machine distorts things," he explained. "Your nipples won't really go out like that." Katye next brought up the profile images, which looked much better. Instead of my breasts having the regrettable ski slope above the nipple (something I never noticed before), now they had the curves of an upside-down cereal bowl.

"You'll do wonderful," said Dr. C.

■ ■ ■ ■

There's nothing like America's consumer culture to convince us that what we have isn't quite good enough. We didn't used to be this way. Americans have traditionally been tough-skinned and self-reliant. At the same time, of course, we've been great re-inventors of the self. Hollywood may celebrate the heroes of the former, but its images reinforce the latter. In breasts, these two strains of character found a new tension by the middle of the last century. Somewhere along the line, lured by Jean Harlow and Jayne Mansfield and the technological promise of postwar America, American women tossed out the make-do-with-what-you-have mentality and embraced a burning desire for outsized nose-cones.

Highly engineered bras helped, but only if you had something to put in them. Kleenex was popular, and so were socks. Falsies, made out of wire, sheet metal, papier-mâché, rubber, cork, elk hair, or cotton, became a multimillion-dollar industry. In its 1951 catalog, Sears offered twenty-two different versions. At that time, surgical solutions to a larger bust were dangerous and rare. Many more breast reductions were performed than breast augmentations. For much of Western history, large breasts were considered a burden and a handicap. Consider the case of poor Elisabeth Trevers, a young Englishwoman who, according to her surgeon, woke up one morning in 1669 "and attempted to turn herself in bed, [but] she was not able . . . Then endeavoring to sit up, the weight of the breasts fastened her to her bed; where she hath layn ever since."

Augmentation came later. Although inserting foreign objects into the body was known to be dangerous, there were always some surgeons and women willing to experiment. The first boob job is attributed to Vincenz Czerny, a Heidelberg physician. He transplanted a benign fatty growth from the backside of a forty-one-year-old singer to her chest in 1895. It was a good idea, since the material came from her own body and was less likely to cause an immune-system rejection, but the result was lumpy and, because the fat liquefied, temporary. That was failure number one.

From that point on, the backstory of implants reads like a horror novel.

In the early twentieth century, implant materials included glass balls, ivory, wood chips, peanut oil, honey, goat's milk, and ox cartilage. What became of the (thankfully few) women who volunteered for these leaps of science? The parable of paraffin offers a glimpse. From the mid-nineteenth century, paraffin injections had been used on facial deformities. Sadly, there was plenty of opportunity; both war and syphilis — which depressed the nose — were great for advancing the art of plastic surgery. Inevitably, the wax was injected into the breast. But by 1920, its limitations were well known. It melted in the sun, for one. It also created lumps and tumors called paraffinomas that eventually had to be excised out, leaving scars. Beyond that, other problems were puss, hardness, blue skin, and feverish rheumatism. At least one woman's infected breasts had to be amputated. As one historian put, the disadvantages of paraffin ranged from aesthetic failure to death.

Of course, women going to dangerous extremes for beauty was hardly new. For a thousand years Chinese women crippled themselves and their daughters to have tiny, deformed feet. Western women literally suffocated while wearing corsets, some of which punctured their internal organs. Women have painted their faces with lead and arsenic and ripped their body hair off with hot wax. Oh, wait, we still do that.

Into this sorry milieu came the plastics revolution and a new breed of unholy implant contenders: Teflon, nylon, and Plexiglas. Several surgeons were moved by the shape of plastic kitchen sponges. In 1957 a Johns Hopkins surgeon implanted a polyvinyl and polyethylene sponge (also made with "foaming agents" and formaldehyde) called the Ivalon into thirty-two women. As one magazine reported at the time, "The material's one drawback is that when it dries inside the breast it becomes a hard lump."

Meanwhile, toiling in a laboratory in Midland, Michigan, chemists were experimenting with different uses for a versatile material called silicone. Corning Glass Works had begun fooling around with the stretchy composite in the 1930s, making it from silicon (an element) left over from its glass production. To this they added organic carbon-based chemicals in various configurations, resulting in a material that was pretty close to miraculous: hardy, inert, and heat resistant, yet soft and flexible. It was a glass-and-plastic hybrid, with the best properties of both. The company thought it might make a good mortar for its trendy glass bricks (they were wrong, but the failed formulation found new life two decades later as Silly Putty). At the beginning of World War II, U.S. Navy officials coveted a similar formulation of silicone, finding it perfect for insulating airplane ignitions (it made long flights to Europe possible) and for lubricating machinery. To guarantee larger supplies of the carbon-based ingredients, Corning partnered with Dow Chemical in 1943 to form a new war-christened giant, Dow Corning, in the American heartland.

When the war ended, Dow Corning was eager for new civilian markets for wartime products. The company began ardently filing patents for silicone polishes and paints, adhesives, silicone shoe rubber (astronaut Neil Armstrong would take a giant step in it in 1969), caulking, and other applications. The medical profession was intrigued by silicone's strength, flexibility, and apparent non-reactivity, and slowly the material made its way into catheters, stents, tubing, and blood bags.

In American-occupied Japan, another, less orthodox use was found for silicone. Drums of the stuff, needed for cooling transformers, went missing from the docks of Yokohama harbor. It turned up in the breasts of Japanese prostitutes, who were being injected with it to better attract enlisted farm boys. The technique spread through eastern Asia and became one of Japan's most popular exports to the United States. But as with paraffin, the industrial caulk-like material was known to migrate throughout the body, form hard lumps, and cause serious infections.

Back in Houston, plastic surgeon Thomas Cronin was holding a new silicone bag of warm blood in St. Joseph Hospital. It was 1959, and the blood bags were a nice change from glass bottles. *My*, he thought, *that feels good. That feels like a breast.*

The era of the boob job was about to arrive.

At first glance, this hard city of oil derricks, pipelines, and banks might seem an unlikely place for such a defining moment in the natural history of breasts. But in addition to its status as the oil and gas capital of the country, Houston in the 1950s was emerging as a major medical hub, in no small part because of the city's oil and gas wealth. MD Anderson Cancer Center had been created in 1941 as part of the University of Texas system. Houston's Texas Medical Center, including several nonprofit hospitals and schools, was well on its way to becoming the largest medical center in the world. At

Baylor College of Medicine, where Cronin worked, a cardiologist named Michael De-Bakey had just pioneered a procedure called patch-graft angioplasty with a Dacron swatch, a celebrated technique still used today. Plastics and chutzpah were revolutionizing medicine.

Add to this a lively burlesque scene, the city's embrace of petro-fueled commerce and technology, and its particular brand of cowboy entrepreneurialism, and Houston was perfect for the Future Boob mantle. Cronin was ambitious, and he'd been thinking about the breast for some time. He was aware of the practice of silicone injections and dismissed it as no good. But when he saw the new blood bags, he reasoned that if the filler substance could be contained in a sac, many of the collateral problems would be solved. He and his chief resident, Frank Gerow, found a receptive audience at Dow Corning. Working with the company, they designed an implant using a silicone rubber bag filled with silicone gel. On the back of the bag, they added several patches of Dacron in the hope that it would bind to the chest wall and keep the sac from ending up in an armpit. Accounts vary about how they tested it. Some authors say they tested it in six dogs, but Dr. Tom Biggs, who was another resident of Cronin's at the time, told me they tested the implant in only one. She was, he recalled, a pound mutt named Esmerelda. When Esmerelda survived the surgery, the doctors called it good. (Esmerelda was not as delighted by her new profile, however. She soon chewed the implant out.)

Next, they needed a human volunteer.

In 1962, Timmie Jean Lindsey was a twenty-nine-year-old woman with a hard life behind her. After her mother died of cancer, she dropped out of high school at the age of fifteen, left home, and married a gas-station attendant. Six kids and twelve years later, she kicked him out for being a slouch and an alcoholic. She then fell hard in love with a steelworker, who talked her into getting a big tattoo. A red rose on her right breast said, "Fred," one on her left breast, "Timmie," and in between bloomed yet another rose. But Fred was a womanizer and things didn't work out. At a checkup, Timmie Jean's doctor audibly gasped when he saw her chest. Feeling ashamed and depressed, she went to Houston's public hospital, Jefferson Davis, for dermabrasion. That's where she met Cronin's chief resident, Frank Gerow. He was another man with a plan for her breasts.

I found Timmie Jean in a small unincorporated town east of Houston. With both Cronin and Gerow dead, she is, on the fiftieth anniversary of her historic implant surgery, the best remaining artifact of the era. Nothing in Houston commemorates the event or the hundreds of millions of dollars that breast implants would soon be pumping into the medical and legal communities. But then again, Houston is not a looking-back kind of place.

"That's how it all started," said Timmie Jean, who's now seventy-nine and, to my jaundiced eyes, surprisingly healthy for having been a surgical guinea pig. A robust and gracious redhead, she works the night shift at a nearby nursing home that no doubt houses a few people considerably younger than she. She welcomed me to the same house in which she has lived for the last fifty years, though the house, much like her chest, has undergone some augmentation, including a couple of small additions to the original shotgun floor plan. Tan with red shutters, it sits not far off Interstate 10, next to a boat-and-generator repair shop and across the street from two large chemical holding tanks. We sat on a couch covered with crocheted afghans in a room crowded with pictures of her children and grand-children. A straw-hat collection decorated one wall, and in the next room, an upside-down pink umbrella served as chandelier above the dining table. Now a widow, Timmie Jean shares the house with her daughter Pamela.

"Unbeknownst to me, implants were in development and they were looking for young women to be the first to have them," she told me in a gravelly Texas twang. "So they brought it up to me. They asked me, would I like to be in a study to have implants? I'd never even dwelled on [my breasts]. I was okay with what I had. After six children I guess they were kind of saggy. I said, 'You know, what I really want is to have my ears pinned back.' My brother had teased me my whole life. They said, 'Yeah, we'll fix your ears too.' "

So in a move that would never pass today's institutional review boards, Timmie Jean got a cosmetic surgery she didn't want in exchange for one she did. She went from a size A or B cup to a size C. "I have to tell you," she said, "they said it would boost my confidence, but I had plenty of confidence." With new breasts and new ears, though, more men did notice her. But there were drawbacks. At the time, she worked in a dress factory, and as a perfect size 12, she was the in-house model. But her new breasts no longer fit into the shirtwaist dresses of the time. And within five or ten years, she said, her implants hardened and sometimes caused shooting pains in her chest. She wasn't able to do aerobics or certain exercises because of the pain. She is self-conscious if anyone hugs her. She has also suffered from rheumatism, and has had two knees and a thumb joint replaced, but she doesn't know if her immune-system troubles were caused by the silicone in her body or by a life of unceasing hard work.

Around the time of her surgery, the doctors asked her if she knew anyone else for their study, and so she recruited her sister-in-law and her sister-in-law's sister-in-law. Over the years, like many women, they also had problems with hardness, pain, ruptures, and symptoms of illness they believed were related to the implants. Her relatives eventually joined a class-action lawsuit against Dow Corning and other makers of silicone implants. But despite her ailments, Timmie Jean never publicly complained about the implants. She even testified before Congress, on Dow Corning's dime, that she was a healthy and pleased customer. One of her daughters went on to get implants, and so did a granddaughter.

Natural breasts have a shelf life. So do fake ones, and it's a lot shorter. Silicone implants, even today, last only ten to twenty years, but, amazingly, Timmie Jean is still walking around with the original specimens. She is a living museum. She knows they've ruptured, because she's been screened, but she doesn't want them removed. "I don't want to go through that," she said. (Surgery to remove implants, known as explantation, can be considerably more involved than the original if it requires cutting away dense scar tissue, calcifications, and hard nodules called siliconomas.) Plus, she said, "I fell on my boobs and they saved me."

Would she do it all again? She's not sure.

"I'd have to look at my options."

For now, she's trying to decide whether or not to return a request from Tom Biggs to examine her. She knows her breasts are of great medical interest. "I suppose I should call him," she said.

"Would you donate your body to science?" I asked.

She laughed. "No, but they can have 'em if they want 'em."

The 1962 enlargement of Timmie Jean launched two cultural tsunamis: a clamor for implants and then, in the 1990s, a clamor against them. Presenting their work to the third International Conference of Plastic Surgery in 1963, Frank Gerow, Cronin's right-hand man, held a cigar and coffee cup in one hand and Dow Corning's Silastic gel breast "prosthesis" in the other. He reflected the beliefs of the audience when he said, "Many women with limited development of the breast are extremely sensitive about it, apparently feeling that they are less womanly and therefore, less attractive. While most such women are satisfied, or at least put up with 'falsies,' probably all of them would be happier if, somehow, they could have a pleasing enlargement from within."

It soon became the fervent stance of the plastic surgery profession that such women were legitimately diseased, either because of "micromastia" — small breasts — or because of their severe psychic inferiority complexes, a handy Freudian concept in vogue at the time. And where there's a disease, there's a cure. One surgeon's autobiography was filled with slump-shouldered, depressive "before" pictures and gleeful, exuberant "after" shots. The message was clear: bigger breasts could change you from a loser to a winner. As recently as 1982, the American Society of Plastic and Reconstructive Surgery told the U.S. Food and Drug Administration that "there is a substantial and enlarging body of medical information and opinion . . . to the effect that these deformities [small breasts] are really a disease which in most patients result in feelings of inadequacy, lack of self-confidence, distortion of body image and a total lack of well-being, therefore due to a lack of self-perceived femininity. The enlargement of the female breast is often very necessary to insure an improved quality of life for the patient."*

For three decades, Cronin and Gerow and their colleagues rushed to fill (as well as to create) the demand for larger breasts. Mastectomy patients represented 20 percent of the total. For them, implants would stand in for what had been brutally cut away with the cancer. For the rest, though, implants promised youth, a certain kind of confi-

* Even recent journal articles, particularly in the plastic surgery field, continue to refer to micromastia as an "abnormal" or deformed condition, despite the fact that small breasts are in fact, perfectly normal. Breast-feeding is usually not a problem as the mammary gland grows considerably during pregnancy.

dence, and lots of attention. The implants came in three sizes, small, medium, and large. The largest was called "the Burlesque." (It's worth noting that at 340 cubic centimeters, it is now merely the average size used in Houston. Implants in the Midwest and East tend to be smaller, as well as less popular.) Gerow reputedly liked big breasts, and it apparently wasn't unusual for him to take a look at an unconscious woman on the operating table in whom he had just placed implants, decide she could handle bigger ones, and redo the whole thing.* With royalties on the devices, the men made a lot of money. So did a lot of other surgeons. One Houston doctor boasted that he could perform as many as seventeen breast augmentations a day. He built a breast-shaped swimming pool for himself, with a Jacuzzi for the nipple. If it were up to me, that would be the site of the nation's implant museum.

By 1985, one hundred thousand women

* According to the seasoned fitting experts at Top Drawer, an esteemed Houston lingerie shop, it's not unusual for women to break down crying because their implants are too big. "They'd tell the doctor they wanted a C cup," said saleswoman Linda Parmley. "Doctors know ccs, not Cs.")

were getting breast augmentations a year, adding some thirteen thousand gallons of silicone gel annually to the nation's mammary capacity. By 1992, two million women had implants, fueling a $450 million industry.

Especially in the beginning, women from the entertainment industry represented an outsized portion of patients. These were also the women clamoring for silicone injections, a practice that continued well into the 1970s and was mostly performed by "cosmeticians." It was cheaper and easier than undergoing implant surgery. From its illicit origins in Japan, the option was popularized in the United States by San Francisco's Carol Doda, credited with being the country's first topless go-go dancer. In 1964, while dancing at the Condor Club, she underwent forty-four injections of silicone, turning herself into an overnight sensation and winning the title of "the new Twin Peaks of San Francisco." It is no understatement to say that Doda changed the landscape of breasts. By 1965, she was appearing in Las Vegas and insuring her mammary assets for $1.5 million. In 1968, Tom Wolfe immortalized her anatomy in *The Pump House Gang:* "Carol Doda's breasts are up there the way one imagines Electra's should have been, two incredible mammiform protrusions, no mere pliable mass of feminine tissues and fats there but living sculpture — viscera spigot — great blown-up aureate morning-glories."

Doda dangled her glories before San Francisco's power brokers while dancing the Swim, the Twist, the Frug, and the Watusi. She did this to live music from a white hydraulic piano that moved up and down. Now that's entertainment. (A word about that piano: it once again made headlines in 1983 when a bouncer shtupping a stripper after hours on said piano was crushed to death after accidentally activating the hydraulic system. The trapped woman waited several hours to be freed by a janitor.)

Thanks to Doda's volcanic success, the pneumatic look became de rigueur for any self-respecting topless dancer, and patrons came to expect it. The strippers instantly saw their tips increase. Corresponding with the popularity of hometown implants, Houston became the strip-club capital of the world. Rick's Cabaret anchored the city's club scene. The club's average bust size was a 38D, according to the *Texas Monthly.* Founded in 1983, it supplied more models to Playboy than any other club. At one point, it was American Express's largest charge customer. Setting up franchises across the country, it went on to become the first publicly traded strip joint. Big breasts were going national.

I looked up Doda, half expecting her to be a tenderloin junkie long dead from some sort of silicone poisoning. But once again, my knockered preconceptions were knocked upside down. Doda went on to live a long and rather fabulous life. Now well into her seventies, she currently owns a lingerie store in a fashionable San Francisco neighborhood and makes occasional appearances with her band, *The Lucky Stiffs.*

Despite the contribution by Doda and her followers to the GDP, federal regulators were not on board with silicone injections. Because the substance was being injected, the FDA classified silicone as a drug in 1965. Alarmed by its poor quality, the agency then prohibited Dow Corning from selling industrial-grade silicone to medical or beauty practitioners and restricted the use of the "medical-grade" stuff to only eight doctors for controlled studies. Even so, an underground trade flourished. An investigation revealed that by 1975, more than twelve thousand women had received injections in Las Vegas alone. There were reports of infection, gangrene, necrosis, and amputations. By 1971, at least four women had died from silicone embolisms, clumps of silicone that had lodged in their lungs or brain. The press covered stories of "Tijuana silicone rot."

While injections were regulated, implants were not. They were classified as a "medical device," not a drug, and the FDA did not have authority to regulate medical devices until 1976. Even then, lobbyists ensured the implants were grandfathered in, meaning they did not need to go through an approval process as long as the manufacturer kept the agency updated on any safety problems. It was an eerie parallel to the sixty-two thousand chemicals also grandfathered in that same year under the new Toxic Substances Control Act (for more on that, see chapter 5). In both cases, the lure and power of technology — technology that would alter women's bodies in wholly unexpected ways — trumped consumer protection.

From the beginning, Cronin and Gerow knew they had some problems on their hands. The sac did not, in fact, prevent the breast from hardening, and the silicone gel proved harder to contain than they'd hoped. The first-generation implants had a ridge-like seam that could be felt on the sides of the breast. Many patients — 41 percent, according to a 1979 study — experienced loss of nipple sensation. An enormous percentage of patients — around 25 to 70 percent by ten years — suffered "capsular contracture" in which the body walled off the implant by creating fibrous scar tissue around it. The scar shell then tightened and shrank, contributing to a visual that became known as "the doorknob effect." If silicone injections came to look like a bag of rocks, the implant resembled one big shriveled stone. A Houston neurologist, who was openly critical of implants, told me he once saw a patient who had been shot. She was a showgirl, and her implants were so hard that the bullet bounced off and saved her life. "They were like doorbells," he said of her breasts.

Doctors speculated the contracture was a response to contamination and infection. Early dissections of the affected tissue revealed that pieces of paper, wood, cotton, talc — essentially pieces of the operating room — had routinely lodged on the implants. The operation itself, initially performed with fairly crude implements like scissors, resulted in a lot of blood and hematomas — gnarly bruises — around the implant. Eventually, surgeons would develop a cleaner, "no-touch" technique with fewer side effects.

Implant makers also attempted to address the problems. Dow Corning created a thinner, seamless bag. That solved the ridge problem, but the bag was so thin the gel freely oozed out of it (this was known as "gel bleed"), and the bags ruptured more easily. Company salesmen were told to wash the leaking implants with soap and water before presenting them to surgeons.

"We had the silicone catastrophe because the implants were made with a not totally impermeable barrier," Biggs, now a retired plastic surgeon in Houston, told me. "It was a bad product."

Beginning in 1982, manufacturers introduced a new, polyurethane-foam-covered implant, called Meme, in hopes it would keep breasts from becoming bulletproof. Patients with these implants did experience lower rates of capsular contracture, and by 1991 this was the most widely used device. But the reason for the success was that the foam was evidently breaking down in the breast, causing a prolonged inflammatory response and "microencapsulations," in which "multidirectional contractile forces cancel one another."

In fact, as the implants got better, they

were actually getting worse. Despite some sporadic earlier testing, it wasn't until 1991 that the FDA released a report that the foam was releasing 2,4-toluenediamine, a known carcinogen. Within days, Bristol-Myers Squibb withdrew Meme from the market. But by then, at least 110,000 women had received those implants. Amazingly, the polyurethane foam being used in the breasts of women was *the same stuff* headed for carpet pads and carburetors. It was never reformulated for medical use, and the manufacturer of the foam was apparently surprised to learn where it was ending up. Many surgeons remember these implants fondly, and foam-covered implants continue to be used in Europe and South America.

(Lest you think the Europeans are making better products, however, a glimpse into the 2011 French implant scandal will set you straight. Jean-Claude Mas, the seventy-two-year-old founder of Poly Implant Prothèse [PIP], is currently facing criminal charges in France for fraud and injury. For nine years, PIP sold implants secretly made with cheap industrial silicone including fuel additives and other chemicals never studied or approved for medical use. The adulterated implants, now installed in a quarter million woman throughout Europe and South America, are believed to be rupturing at higher than expected rates and causing inflammation.)

By the time of the foam revelations in the United States in 1991, another, bigger wave was crashing. Patients whose silicone implants had ruptured were reporting various idiopathic illnesses, everything ranging from fatigue to joint pain to lupus. The stories were all over the media, in the hearing rooms of Congress, and in the mailroom of the FDA. Several multimillion-dollar judgments were awarded to individual patients in juried courtrooms. The agency's commissioner at the time, David Kessler, stated that "we know more about the life span of automobile tires than we do about the longevity of breast implants."

In 1992, the FDA issued a moratorium on silicone implants except in women following breast cancer surgery who agreed to participate in clinical studies. (Saline implants were still available.) By 1995, half a million women were suing the makers of implants and their surgeons. Faced with 20,000 lawsuits and 410,000 impending claims, Dow Corning declared bankruptcy. The company eventually entered into a $3.2 billion settlement with 170,000 women. It was the largest-ever class-action settlement at the time.

A recent college graduate, I clicked into the implant controversy around the time of the moratorium. I'd studied pesticide contamination of farm workers, chemical-plant explosions in Bhopal, and radiation poisoning in Chernobyl. Then an Exxon tanker cracked open in Valdez, Alaska, leaving birds slickened and imperiled. When I saw numerous news reports that plastic implants were rupturing inside women, who then came down with mysterious immune-system ailments, it seemed to make perfect sense as another example of corporate malfeasance and crimes against nature.

But now, after twenty years of study, science has not backed up most of these claims. Research to date has found that women with silicone implants, even the older versions, do not have more immune-system diseases than their *au naturel* peers. Some research suggests that they have slightly higher rates of immune-related symptoms such as fatigue and arthritis, but other studies contradict that. In 2011, the FDA reported that implant patients have higher rates of a very rare cancer called anaplastic large-cell lymphoma. This cancer grows in the cells of the scar tissue surrounding the implant, but it is distinct from breast cancer. Implant patients do not have statistically higher rates of breast cancer, but they do have higher rates of lung and brain cancer. This is possibly because of migrating silicone, but more likely due to associated lifestyle factors such as smoking.

To be sure, many things are still distressing about the implant story. Doctors and manufacturers profited by introducing a poorly understood substance into women's bodies. Implants were not engineered well at the beginning, they were inadequately tested, and patients were not always informed of the many real risks of the surgery, or of the high failure rates of the devices. Ultimately, the immune-system scare became a distraction from these other issues. As the legal scholar Julie Spanbauer put it in 1997, "The message that never reaches the public is that the majority of women with breast implants, those who received their implants before approximately 1992, have become nonconsenting, de facto participants in these and, unfortunately, in future safety studies." But the crazy thing about the implant story is that no one comes out looking clean. Everyone was out to exploit everyone else, the media included. The implant patients themselves proved able opportunists; if the medical studies are correct, many more women jumped on the class-action gravy train than had claim to do so.

After 1992, the numbers of women receiving implants in the United States briefly plummeted from a high of 150,000 to about 30,000 annually during the conditional moratorium. But by 2007, a year after the FDA approved the next generation of silicone implants — still based on the original Cronin-Gerow concept of gel in a baggie — that figure had increased nearly 1,000 percent. Despite the recession, the worldwide market for breast implants is roughly $820 million a year and growing at 8 percent a year. Between five and ten million women are walking around with implants.

After a fourteen-year silicone hiatus, the fake-boob industrial complex was fully back in business.

Before I left Houston, I was invited to observe Dr. Ciaravino in action. I was curious to know who his patients were and, I suppose, to bear witness to this visceral alteration of the natural breast. I steeled myself for the operating room by watching stills and YouTube videos of implant operations. One set of photos showed how the nipples were cut open and tubes placed in them to fill saline sacs. It gave the term *breast-feeding* a whole new visual. This is nurture turned on its head, a gut-churning reversal of lactation.

Fortunately, I would be spared that. Ciaravino prefers to work through a small, neat incision under the breast in what's called the inframammary fold. He has a kind and compassionate staff, and he uses an experienced anesthesiologist. If you want implants, this seems to be a good place to come. Ciaravino is known for what he calls a "no-bleed" surgery. He cleaves the chest muscle from the rib cage using a pen-sized cauterizing tool that seals the tissue as it goes. He doesn't want blood because a dry seal around the implant reduces risk of encapsulation and because iron is a primo nutrient for bacteria.

Katye had assured me that implants "are the most studied medical device in history," and that they are now "100 percent safe." But are they? The truth is that ongoing studies continue to raise basic health questions, and the FDA and even implant makers acknowledge as much. When the agency approved the new silicone implants in 2006, it was under the condition that manufacturers carry out ten-year follow-up studies. In its fifty-two-page product insert data sheet for MemoryGel implants, the Mentor corporation summarizes the results of the first three years of this study. In addition to finding an eye-raising three-year complication rate of 36 to 50 percent for implant patients (including trickier cancer reconstruction patients), the insert states, "Compared to before having the implants, significant increases were found [in patients] for fatigue, exhaustion, joint swelling, joint pain, numbness of hands, frequent muscle cramps, and the combined categories of fatigue, pain, and fibromyalgia-like symptoms. . . . These increases were not found to be related to simply getting older over time."

The Mentor study has found that the three-year reoperation rate for patients (depending on whether they received augmentation or reconstruction procedures) is between 15 and 29 percent. Some reoperations are done because of cosmetic failure. The door-knob effect isn't the only visual problem. A perusal of "bad boob jobs" on the Internet yields a grim parade of Uniboob (also called "bread-loafing") in which the implants migrate toward each other; Double Bubble or "bottoming out," in which the implant drops below the breast fold, creating what looks like a double-decker breast as well as serious asymmetry; Highballing, in which the implant sits too high; and various degrees of wrinkling and dimpling. A woman buying implants today may still be buying into recurrent surgeries, costly, regular MRI screenings (to detect "silent" ruptures), and a reduced ability to detect early breast cancer (the implants can block effective mammography).

Even more troubling, though, is that some women continue to report problems with nipple sensation and breast-feeding because nerves can be damaged during surgery. While Ciaravino says these effects in his practice are rare, a major review of the literature from the Institute of Medicine in 2000 stated that women with either silicone gel–filled or saline-filled breast implants showed lactation insufficiency (not enough milk) at rates ranging from 28 to 64 percent. The FDA's Breast Implant Consumer Handbook further states, "It is not known if a small amount of silicone may pass from the silicone shell of an implant into breast milk. If this occurs, it is not known what effect it may have on the nursing infant."

Perhaps few women with implants are interested in nursing, but one would expect (and hope) that a great number are interested in sexual sensation. Let's be clear about what these under-sung side effects mean: in a world where breasts are considered purely sexy, we jeopardize the central natural functioning of breasts (lactation and dynamite neural sensation) so that they can be even more sexy, to the point where the improvement actually eliminates the sexual feeling in this allegedly sexy organ. Now we can have hard, lifeless replicas of something sexy. Plastic surgeons understand this: in 1976, a pair of them observed in a trade journal, "Fortunately, patients undergoing plastic surgery of the breast are concerned more with getting rid of a deformity and achieving a desired body image than with maintaining or improving mammary sensation."

They appear to be correct. When I learned about the continuing problems attributed to implants, what I was most astonished by is that so many women still want them, especially in Texas. Even with their contractures and ruptures and door-knob breasts and lifeless nipples, most women say they are *happy* with their implants. At least over the short term, many implantees report increases in self-esteem and sexual self-confidence, if not actual sensation. As the brain is the body's largest sex organ, this makes some kind of sense. Mentor's study to date shows that of 456 new augmentation patients within three years of surgery, 98 percent would have the procedure again. Other studies show that even seasoned patients do it all over again after their old implants have crumpled and died.

Do big breasts really have that much more fun? Or are we, as critics like Naomi Wolf suggest, hopelessly brainwashed by a beauty myth designed to keep our minds distracted by frivolity? That women should feel good about themselves is their right; but that they should feel so bad about themselves in the first place shows that the modern boob job represents a great failure of the imagination.

How do we convince our daughters not to join the legion of women who feel they have such limited avenues to happiness? Unfortunately, that challenge seems to be only growing. Double-D breasts on skinny women are not all that common in nature. (Barbie's proportions are naturally found in one out of one hundred thousand women, according to researchers from the University of South Australia; Ken's bod, by contrast, is found in one in fifty men.) Big, fake breasts have so thoroughly saturated mainstream entertainment and media that they've created a new standard by which boys judge girls and girls judge themselves.

Thanks to the alliance of two kinds of silica-based technologies — breasts and computer chips — most young people learn about bodies and sex from the Internet; they have seen many more factory-made breasts than real ones. In this crowd anyway, natural breasts just keep losing traction.

The patients in Dr. C's surgical suite knew as much. The first patient I met was a twenty-nine-year-old named Gloria who weighs ninety-nine pounds. A recent college graduate, she has a two-year-old. When Dr. C came in for a pre-op review, she took off her robe to reveal breasts that are quite fantastic: firm and nicely rounded, probably a B cup. A delicate butterfly tattoo lay between them, and her back sported a geisha surrounded by cascading pink blossoms. She would be getting 275-cubic-centimeter implants, for "a full C." I asked her what made her decide to do this. "I just want to put back what I had before my son was born," she said.

Gloria was, explained Ciaravino later, ideal for surgery. "You want the little skinny ones," he said, reflecting a truism among all surgeons. "The ideal patient, she's had a couple of kids, she's good looking to start out with. If you have funky breasts to begin with, they're going to be funky after. It's not going to be picture perfect. It's all a relative improvement."

As Gloria went off to meet the anesthesiologist, Dr. C strode to OR1 to place saline implants in a forty-one-year-old Filipina nurse. I watched as he rolled up the silicone shell like a pirouette cookie and then pushed it through the incision. After he flattened it out with his fingers inside her breast, he connected some tubing to its valve. He talked as he went. "With silicone, you have to push it in. You can get a little wear and tear. And you can see her filling up here before your eyes. She's blowing up like we're filling up a balloon here." A machine was pumping in 340 cubic centimeters of saline fluid through the tube and into her breast. When she was fully inflated, Ciaravino stuffed some escaping yellow breast tissue and fat back into the incision, and then he and a nurse took turns sewing her up. She moaned and thrashed a bit. The anesthesiologist by her head adjusted her dose.

Next we went to have a pre-op with Courtney. As always, Ciaravino had a half-dozen patients rolling through the surgical suite in a conveyor belt of calm efficiency.

Courtney was sitting with her husband, who wore a T-shirt and baseball cap. They were both in their mid-twenties, from a small town not far away. Courtney, a former cheerleader, now owns a tanning salon. She has dark shoulder-length hair and strenuously plucked brows. A mother of two small kids, she was eager for a better body. In a Texas accent, she explained what brought her here. "I'm a very small B or A, probably A. When I was in high school, I probably had a small C. My sister, she didn't have big boobs, but after her daughter was born, hers stayed big. But mine, they got smaller every time. My sister-in-law used him," she said of Dr. C. "Some of my good friends and customers have implants. I just want a full C. I don't have to look good for anybody else, just for myself, I guess to help me feel better about myself."

When I next saw Courtney, she was lying asleep on the surgery table. Her uncovered breasts had been marked with a blue pen. The ink made dotted lines along the contours of her flesh like rivers on a topographical map. Her torso had an orange-tinted tan, and her breasts were indeed small. She'd be getting 350-cubic-centimeter Mentor MemoryGel implants, bigger than the old-fashioned Burlesque. Dr. C explained that Courtney was a bit of a challenging case. "The issues here," he said, "are that she doesn't really have a defined fold at all so we're going to have to sort of create that. It puts her at a little bit of a risk for the implant moving down. So when I close it I'm going to have to tack it in place." He went to work on her, cutting and singeing below her pectoral muscle. He pulled the space open with an instrument resembling a shoehorn. It's called a Biggs Retractor, named after the Houston surgeon who trained with Cronin. Ciaravino motioned me over to catch a glimpse of Courtney's heart beating between her ribs. The wound smelled of burning flesh.

When I'd seen enough, I stepped out to chat with the next patient in line, an insurance agent named Katie. She was thirty, a brown-haired mother of two, from Orange, Texas. She said she's never been under anesthesia before, and she's nervous. She wanted to go from a size A to just a small C because "we're conservative people," Katie explained. "It's not like all my life, I said, 'Oh my God, I want boobs, I want fake boobs.' I just want my clothes to fit better without having to buy an extra padded bra." Not having implants, she said, "I'm like the minority, I think, in our circle of friends."

She laughs, patting down her hospital gown. "It's peer pressure."

DDT on Jones Beach, 1948

5
Toxic Assets:
The Growing Breast

I tell people I come from a different planet because the planet I arrived on is so unlike the planet of the 21st century. There were no plastics; there was less carbon dioxide. There were more fish in the sea. I come from the pre-Plasticozoic era.

— SYLVIA EARLE,
National Geographic explorer-in-residence

The same year that Timmie Jean was exchanging an ear tuck for a boob job, Rachel Carson published a book about the destructive power of pesticides. These two events had more in common than it might appear, for both heralded a new era of synthetic compounds that would forever alter breasts. In 1958, the nature writer and biologist had received a disturbing letter from Olga Huckins, a gardener in Duxbury, Massachusetts. It described how the local authorities had sprayed fuel oil and DDT (dichlorodiphenyltrichloroethane) to kill mosquitos, leaving scores of songbirds dead in her neighborhood and in her very backyard. Huckins wrote that birds fell from the sky. Others perished in grotesque postures around her birdbath, their claws splayed, their bills gaping open.

Carson was already known as a voice for nature. The first female biologist hired by the U.S. Bureau of Fisheries, she'd written several odes to the sea, including the wildly popular best seller *The Sea around Us.* It won a National Book Award in 1952.

Inspired by Olga's letter, *Silent Spring* was a measured and eloquent argument against the indiscriminate use of synthetic pesticides, which Rachel Carson called "elixirs of death." She described the little-known work of scientists showing how DDT and its ilk caused damage beyond their target insects, affecting birds, fish, and other vertebrates. She got a couple of things wrong in the book, such as the statement that few carcinogens exist in nature (in fact, there are many, including the sun, wood smoke, and numerous viruses and fungi). She could be melodramatic, describing a future lifeless world and quoting Keats ("The sedge is wither'd from the lake, / And no birds sing"). But history would prove her correct about the unanticipated effects of widely used neurotoxins in the environment. She introduced a nation to the idea that human actions and the natural world were inextricably linked, and that people had some obligation to protect that world. The United States' seminal environmental legislation of the 1970s can be traced back to the wide constituency she built.

Carson made *ecology* a household word. Moreover, she placed the human body squarely within that ecology. She described the rising rates of cancer since World War II, which launched a new era of synthetic organic chemicals. She pointed out that we are now living with persistent industrial chemicals in a way our ancestors never did. "For the first time in the history of the world, every human being is now subjected to contact with dangerous chemicals, from the moment of conception until death," she wrote. Or as The Who put it in 1966, "I was born with a plastic spoon in my mouth."

Carson didn't know that many of these compounds have the ability to alter human hormone systems, but she presciently described chemicals accumulating in the sex organs of birds and mammals and corresponding drops in sperm levels. She was struck by reports that roosters were losing their wattles and that sperm counts were found to be low in pilots who sprayed pesticides from the sky.

The term *endocrine disruptor* wouldn't be coined for another thirty years, when a concerned group of wildlife biologists gathered to exchange research on everything from intersex fish to birds that refused to act like parents. To the scientists, the evidence was mounting that synthetic compounds in polluted areas, notably the Great Lakes, were altering the cells, bodies, and behavior of these animals in ways not seen before.

Until that time, most people thought synthetic estrogens showed up only where we intended them to, namely, in medication. But hormones are famously sneaky. They do their work in our bodies in infinitesimally small quantities. One molecule of a hormone fits into one receptor on a cell like a key in a lock, unleashing a chain of biochemical events. These are the keys that govern everything from cell division to metabolism to hair and breast growth to cognitive performance on spatial tasks. Hormonal changes during women's monthly cycles affect how they smell things, how they perceive the faces and bodies of the opposite sex, and even how they think.

Some studies show that when estrogen levels peak mid–menstrual cycle, women reportedly get better at verbal and fine-motor skills.

If a foreign imposter shows up on the receptor, however, the body's responses become all but impossible to predict. Some foreign estrogens, called xenoestrogens, occupy the receptors so native estrogens can't do their jobs. Others pass the security test and turn on estrogenic responses. Some seem to mess with the body's feedback loops, causing the brain to release more or less hormone than it otherwise would.

It is now known that some imposters occur naturally, like estrogenic substances in plants. Why would plants bother to make estrogen mimics? For the same reason they make acids, poisons, and thorns. Plants are smart, or at least evolutionarily successful. One jaunty example is marijuana, which figured out how to make THC (tetrahydrocannabinol), a compound that happens to fit perfectly into pleasure receptors in the human brain, all but ensuring its spread around the world. Which begs the question, do humans cultivate marijuana, or is the plant actually cultivating its human growers? Marijuana also releases compounds that inhibit testosterone in people who consume it. Long-term male stoners are known to sometimes grow small breasts, and they also face increased risk of breast cancer. Is this a coincidence, or is the plant somehow benefiting from having less aggressive users who giggle easily and croon folk songs?

A couple of dozen plants are known to produce high levels of phytoestrogens, which essentially act as oral contraceptives, in order to knock back their predators. When sheep eat a particular strain of clover, they can't reproduce. Humans have long taken advantage of these plant properties, using certain herbs and fruits to prevent pregnancy or induce miscarriage. Hippocrates knew that the seed of Queen Anne's lace, when eaten, worked as a contraceptive and morning-after pill, as did an infusion of pennyroyal. Giant fennel, found by the Greeks in the seventh century BC, was so coveted for the contraceptive trade that it was harvested to extinction.

All of this was far from the mind of Patricia Hunt in 1998, when she was working in her lab at Case Western Reserve University in Ohio. She is an experimental biologist with a fascination for aneuploidy, the chromosomal abnormalities that cause miscarriages or birth defects (like an extra copy of chromosome 21, which causes Down syndrome). These are the things that can go wrong in the earliest moments of conception and fetal development. Humans are supposed to have twenty-three pairs of chromosomes, but sometimes there's a little, or big, miscalculation, especially in the eggs of older women. Hunt wanted to know why. She hoped studying mice would yield clues.

As was typical, she was working with mutant mice and some control, or normal, mice. One day, she noticed that her control mice, which were supposed to have healthy eggs, were producing a high rate — 40 percent — of abnormal eggs, when the usual rate is 1 to 2 percent. Something had gone terribly wrong. "We checked everything," she told me, "the air in the facility, were there pesticides coming in? It took us weeks. We finally noticed some wear and tear on the animals' plastic cages." It turned out a temporary worker was using the wrong detergent to clean the cages, and it was degrading the cage walls, causing a substance called bisphenol A to leach into the mice's food and water.

Bisphenol A, or BPA, was an artificial estrogen developed in the 1930s as a possible drug to prevent miscarriages in women. It didn't work for that, but the compound soon found other uses, like making polycarbonate plastic. BPA's molecular structure is simple and elegant: two joined hexagons, each made up of six bonded carbons. When these are neatly stacked to make a long-chain polymer, the material is incredibly strong. You can drive over a polycarbonate water bottle and it won't break. It's also cheap, derived as a by-product of refining petroleum. Unfortunately, BPA's versatile ring structure bears a close resemblance to estrogen. Now produced in mind-boggling quantities of two million pounds per year in the United States, for annual profits in excess of $6 billion, it's in everything from the lining of canned food to dental sealants, to consumer products like CDs, cellphones, and bike helmets, to the shiny paper receipts we get from the grocery store. It was also in lab equipment at Case Western Reserve.

"I don't think anybody in the aneuploidy field really believed in environmental effects. I know I didn't," said Hunt, who's petite and trim, with short, tousled hair. "That fad came and went in the 1970s." That was when, spurred by the connections suggested in *Silent Spring,* scientists realized that it

146

was much harder to attribute disease to pollution than anyone expected. "We knew that maternal age was the huge, main event [for abnormal eggs], and that trying to see any other effect would be like trying to see a snowball fight in the midst of an avalanche. That's why this really caught our attention," she said.

Another, much more personal, endocrine effect caught her attention five years after the leaching-cage problem. Hunt was diagnosed with breast cancer. She knew she was a so-called DES daughter (her mother had taken diethylstilbestrol while pregnant) ever since she'd received abnormal pap-smear results in college. In other words, in the earliest beginnings of her own life, she was exposed to a potent hormone that wasn't supposed to be there. Ironically, the drug that replaced BPA in the 1930s to prevent miscarriage was DES, a more powerful estrogen (but alas, no better in hindsight at preventing miscarriage). Because of their early-life exposures, some DES daughters are known to suffer from rare, devastating reproductive cancers, malformed uteri, and higher-than-normal rates of breast cancer. When they've been able to bear children, their daughters are also more prone to breast cancer. DES sons have lower sperm counts and higher rates of genital birth defects. DES was not discontinued in pregnant women until 1971, nearly three decades after it was introduced. An estimated five million people were exposed through pregnancy, and many millions more consumed DES-tainted beef and poultry. Because of DES, scientists learned that chemicals can and do cross the placenta, which they once didn't believe was possible.

Hunt has described these events as being twice struck by lightning. Two artificial estrogens were now determining her destiny. She's since devoted her career to studying the strange and ominous effects of BPA in her mice. Hunt and numerous other researchers around the world have found that early-in-life exposures to BPA can cause early puberty, lower sperm counts, changes in mating behavior, predisposition to obesity, increased rates of breast and prostate cancers, and increased rates of miscarriages. All these troubles were observed in rodents, many of which were exposed while in their mothers' wombs.

"The long-term ability to reproduce is my focus," she explained from her immaculate lab, now housed in a spanking new building at Washington State University in Pullman. Her office is filled with "chromosome art," prints and ceramics showing the spindly wormlike forms of the substance that begins all life. Her husband's office down the hall sports a wooden lamp with a sperm-shaped pull cord.

Hunt showed me the basement "vivarium" where her strains of mice live, the breeders, the studs, and the babies. She no longer uses cages made with BPA. She saves that for her experiments.

When she doses her mice with BPA, she sees abnormal behavior or cells in the mother, the pup, and, later, the pup's children, for what she terms a "grandmaternal effect." One dose effectively damages three generations, just as some researchers are seeing with DES and humans. Hunt's studies are controversial because sometimes the results haven't been replicated. Hunt has a reasonable explanation for this: the effects depend precisely on when the mice are exposed. "The developing fetus is exquisitely sensitive to environmental factors," said Hunt. "There are critical windows, sometimes just one or two days long, in which a tiny dose of chemical can send the wrong message to cells, and other days when the window has shut and the mouse will develop normally."

Depending on that timing, in the mama

mouse ovaries she often sees eggs in which the chromosomes aren't lining up right. Normally, these eggs shouldn't even be viable, but for some reason, the body's quality-control checks aren't working to destroy them. "We want to see the earliest possible defects, and here it is, *bam*. It means we can see things in the early stages of making an egg that should lead to mistakes, and they do." She showed me a magnified image of scattered red wormlike chromosomes, the confounding mistakes in germ-cell scaffolding caused by BPA — a "funny little beast," as she calls it. Pointing to the worm pattern, she noted, "The chromosomes are completely disorganized. I wouldn't want my eggs looking like that spaghetti."

Looking at the images, I could almost feel my ovaries twisting up. But these were mouse eggs. Did her work have relevance for us?

"That [multigenerational effect] for me was huge," said Hunt. "It would be a fifty-year problem in humans. We think these effects are at doses that are environmentally relevant." She was saying that people today are exposed to similar levels per body weight as her mice. "How would we see this in humans?" she asked. "Increasing rates of miscarriages, low sperm counts, testicular problems. We already do see that. How long will it be before we say, 'Holy cow, are we crippling both sexes at once?' " Hunt is frustrated — passionate, even — that U.S. and many international agencies still have not taken a strong stance against BPA. (As of this writing, though, ten states have banned it for use in baby products. France has banned the substance from water bottles, and Denmark has banned it from food packaging meant for children under age three.) She reminded me of the once-staid climate scientists who have become some of the most outspoken critics of U.S. climate and energy policy. They know too much to keep quiet.

In the meantime, no one in her lab drinks out of a plastic bottle.

A couple of Pat Hunt's observations have big implications for breasts. The first is that diseases like cancer can start in the womb, even in the egg, in its earliest beginnings. The second is that synthetic chemicals that whack out our hormones are popping up in wholly unexpected places, like our grocery receipts. Since her first accidental discovery in the lab, hundreds of other scientists have also studied BPA, wondering, among other things, what it does to how the breast grows and develops.

Numerous studies have confirmed that BPA activates the estrogen receptors on breast cells and can cause cancer cells to replicate in a dish. In addition, BPA has been shown to cause normal breast cells to act like cancer cells, growing invasively when they're not supposed to. BPA and similar compounds can turn on and off genes in breast cells in ways that foster cancer. When fetal or young rats are fed BPA, they can be more susceptible to getting cancer later on when they are exposed to other carcinogens. There is something about BPA that alters the way rats' mammary glands grow and respond to later threats. In other rat experiments, prenatal exposure to low doses of BPA caused lesions in mammary glands. When researchers at Yale fed DES (a very strong estrogen) to some pregnant mice and BPA (a weak estrogen) to others at levels approximating those found in pregnant women, both offsprings' mammary glands produced more of a protein called EZH2. Higher EZH2 levels are associated with an increased risk of breast cancer in humans. And it didn't matter much to the glands whether they received the strong or weak estrogen. This was a chilling discovery for those toxicologists still clinging to the ages-old adage that the dose makes the poison.

Cancer isn't the only concern. There is also some evidence that chemical exposures, delivered very early in life, can alter the way the milk-making structures grow, as well as re-jigger the timing of breast growth during puberty. In industrial countries, human breasts are appearing earlier in childhood (stay tuned for the next chapter), and women in the United States suffer from more problems breast-feeding than women in Africa or South America.

It's treacherous drawing connections between lab rodents and people, but if our historical DES experience is any guide, the lab data can be telling. Our mammary glands are similar to those of rodents, and we are sensitive to the same hormones. These potent chemical signals are the ancient currency all vertebrates share. Estrogen is the oldest steroid of them all, with our cellular receptors originating over 550 million years ago. Our bodies inherited an eons-old ability to absorb it from our surroundings, even after we learned how to make it ourselves. We are a modern motif on an ancient device.

Why are our estrogen receptors such

disappointing judges of character? Why are they so promiscuous, hooking up with many different kinds of compounds? Two reasons. One is a central theme of this book, so it's worth saying again: organisms like us are designed to be biologically responsive to the world around us. Scientists call this "pheno-typic plasticity." It means we can tweak our bodies a little bit if it looks like we need to, *regardless of our DNA.* Our hormone recep-tors are sensors; they are one of the ways we gauge the world. The estrogen receptor, though, is confoundingly catholic in its tastes, more so than the other steroid recep-tors. Its lock will take any key with a carbon ring structure, and others besides. As bad luck would have it, this structure is the same building block used to make numerous synthetic compounds, including many plas-tics, solvents, and pesticides.

Our breasts, for better or worse, have more varied and more sensitive hormone receptors than other organs. If anything, their receptors veer toward oversensitivity. They need to sense the environment to store fat and grow at the optimal time and feed an infant in an uncertain world. What might have been handy once upon a time now looks increasingly like a liability.

This leads us to the second reason, which

is that our receptors never before had to dance with so many bad actors. In the past, estrogen mimics were mostly plant estrogens, and many of them are weak and possibly even beneficial. Some phytoestrogens block our receptors in a good way. Some studies suggest that high-soy diets, for example, help prevent a recurrence of breast cancer in survivors, as well as delay breast development in girls, perhaps by protecting their cells from their bodies' more active estrogens. We coevolved with plant estrogens, and that's probably why we can metabolize these compounds quickly. We were also exposed to them seasonally and in limited amounts. BPA, on the other hand, comes to us in a daily drip and, notably, reaches us in the womb.

BPA was proved eighty years ago to act as an estrogen. You might wonder, as I did, why on earth, if BPA was known to behave estrogenically, did it go on to become the principle ingredient in ubiquitous polycarbonate plastic, an ingredient known to break down easily and escape into the environment?

The answer to that would be a long and sorry tale about the lack of government oversight, inadequate testing (in fact, virtually nonexistent testing) for hormonal effects in safety studies, and the phenomenal power of the chemical and pharmaceutical industries — à la Big Tobacco — to sow seeds of scientific doubt and maintain a favorable regulatory landscape.

If known estrogenic compounds such as DES and BPA could fly so long under the safety radar, what about the other thousands of chemicals in our world, especially the ones that are not obvious hormone disruptors? How safe are they? In the United States, seven hundred new chemicals come to market each year, joining the eighty-two thousand already in use. Of these eighty-two thousand, only a few hundred have ever been tested for health effects. Despite the thirty-five-year existence of government regulatory agencies and their guiding laws in the United States, including the Toxic Substances Control Act, only five chemicals have ever been banned. DES was still manufactured in this country as recently as 1997. Unlike in Europe, American companies are not required to perform safety studies on chemicals before they introduce them into the marketplace. In fact, they have a strong incentive not to perform them. In the United States, every chemical is assumed safe until proved guilty. The burden to do that falls on government and university scientists, who don't have the institutional muscle or resources to keep up. It takes years of work to prove that a chemical causes harm, and a shield of proprietary industrial secrets has kept manufacturers from revealing which chemicals they are even using. These recipes may be private, but they swim in your bloodstream and in mine. Of the 650 top-volume chemicals in use, four billion pounds get released into the American water and air each year. Forty-two billion pounds are made here or imported *each day* for use in products and materials.

The U.S. government does occasionally single out some chemicals for lab-animal testing, and the Environmental Protection Agency has recently finalized new tests for routing out the top suspected endocrine disruptors. The official tests, however, focus on certain well-established endpoints: the liver, the kidney, the genitals, and the brain. "They leave the mammary gland in the trash can," pointed out Ruthann Rudel, a toxicologist at the Silent Spring Institute, a Massachusetts-based nonprofit that advocates stronger environmental testing.

The mammary gland, once again, occupies a forlorn and forgotten place in science. This is especially disturbing since independent scientists have found that the mammary gland is the *most sensitive organ* to known culprits such as BPA, DDT, and a common weed-killer called atrazine. Banned in Europe but used in the United States to the tune of seventy-five million pounds a year, atrazine is a major contaminant of drinking-water supplies. In the body it appears to increase the activity of aromatase, an enzyme that converts testosterone and other hormones to estrogen. In affected frogs and fish, scientists are seeing damaged reproductive organs. Rats who are exposed early in life are showing signs of altered mammary gland development and increased incidence of mammary tumors.

The journal *Cancer* reported in 2007 that limited testing (mostly by independent scientists) has so far found 216 chemicals known to cause mammary gland tumors in animal studies. These include chlorinated solvents, products of combustion, pesticides, dyes, radiation, disinfectants, pharmaceuticals, and hormones. Twenty-nine are produced at levels greater than one million pounds per year in the United States. Seventy-three are present in consumer products or in food. Roughly one thousand chemicals have so far been shown to alter animal endocrine systems. These compounds include anti-androgens and thyroid impersonators, not just estrogens. One of Rudel's projects is to corral various laboratories and agencies to develop standard tests for recognizing damage to the developing mammary gland.

We've genetically modified our crops to be able to protect them from the ill effects of pesticides, but we haven't yet figured out how to modify our breasts or livers or brains. Perhaps someday we'll have GMO breasts along with GMO corn. In the meantime, it's next to impossible to draw a line between a particular chemical in our lives and a disease or symptom that might show up decades later. When we do notice a problem with a common drug or chemical, it's because the resulting effects are either very unusual (rare vaginal cancers in DES daughters, mesothelioma in asbestos workers) or very obvious (shortened limbs in children whose mothers took thalidomide to prevent morning sickness).

With DDT, the effects have been more subtle and more varied. But the eventual catalogue of ills, recounted by Rachel Carson and many others, made the endocrine-disruption hypothesis possible.

A chlorinated hydrocarbon, DDT had a

great run during World War II as an insecticide, sparing soldiers from grim deaths by typhus and malaria. After the war, its manufacturers were eager to grease its transition to civilian uses. As Brigadier General James Simmons rhapsodized, "The possibilities of DDT are sufficient to stir the most sluggish imagination. In my opinion, it is the War's greatest contribution to the future of the world." It was sprayed from airplanes over fields, suburbs, playgrounds, and country roads. It was aggressively marketed to housewives and farmers alike. By 1950, it was so widely used that 100 percent of houseflies were resistant to it. Still, its use continued unabated, and by the early 1970s, 1.3 trillion pounds had been sprinkled, sprayed, and dusted in the United States. Many adults today remember gleefully running or biking through the fog sprayed by DDT trucks and planes in the 1950s and 1960s.

After dead insects, birds were DDT's first notable victims. Most famously, the pesticide wiped out populations of bald eagles, peregrine falcons, and brown pelicans because it weakened their eggshells and altered nesting behavior. It wasn't long before scientists started looking for effects in other animals and humans. They measured it in human blood, umbilical cord blood, and breast milk, and they found associations between high levels of the pesticide and the incidence of diabetes, miscarriage, poor semen quality, shortened duration of lactation, preterm birth, low birth weight, and decreased cognitive skills in children.

DDT and its main breakdown product, DDE, are very difficult to get rid off. Fat-loving, stable molecules, they reside in animal tissues (as well as in soils and streams) and stay there for decades. Thirty years after the ban, these molecules are still detected in most human blood samples tested by the Centers for Disease Control and Prevention.

Several studies have compared the DDT levels in women with breast cancer to those in women without breast cancer. Nothing remarkable showed up. But then in 2007 researchers looked at blood samples collected from mothers in the 1960s. The younger women, the ones exposed to the most DDT as *girls* (at levels still measurable in their blood because of DDT's long half-life), had a fivefold increase in breast cancer incidence later in life. The girls who were older than fourteen when DDT was introduced had no increase in the incidence of cancer. For the girls who received their hits in early youth, the blow was much worse. In other words, the timing of the exposure was critical, similar to the effects seen of radiation in bomb survivors and BPA in mice. This study may well help explain why women born after 1940 have much higher levels of breast cancer than women born before. There were simply more chemicals soaking their childhoods.

DDT was just the first in a line of troubling organic pesticides. Next came dieldrin, aldrin, chlordane, heptachlor, hexachlorobenzene, and others. After World War II, our thrumming petroleum and chemical-weapons machinery readjusted beautifully to civilian outlets. Cheap by-products of fossil-fuel production found myriad new uses. Benzene became a building block for BPA, flame-retardants, insecticides, and scores of other formulations. Today we use thirty times more synthetic pesticides than in 1950, nearly nine pounds per person. When Rachel Carson wrote *Silent Spring,* she noted that one out of four Americans would get cancer in their lifetimes. Now, the rate is 1 out of 2.5. An aging population accounts for much of that change, so it helps to look at age-adjusted rates for specific cancers. When we do, we see that the incidence is notably rising for endocrine-related cancers such as breast, prostate, and thyroid. We now know many chemicals have the potential to cause cancer and much more subtle effects, as we'll see. It has taken three generations to find out.

Chemical World News reacted to the publication of *Silent Spring* by calling it "science fiction, to be read in the same way that the TV program, *'The Twilight Zone'* is to be watched." But the public responded to Rachel Carson, and by the mid-1970s, Congress had passed laws to begin to regulate the chemical industry. In 1972, DDT was banned.

Carson wrote, "If we are going to live so intimately with these chemicals — eating and drinking them, taking them into the very marrow of our bones — we had better know something about their nature and their power."

She'd no doubt be discouraged by the extent to which chemicals are even more embedded in our food systems and personal lives than they were fifty years ago. But she'd be fascinated by our increased understanding of how they affect so many bodily processes.

Eighteen months after *Silent Spring* was

published, Carson was dead at age fifty-six. She had breast cancer.

BEGINNERS' LUCK

The teen-ish set loves these Gossard bras because they fit to start with and grow as you grow. Tucks let out! And these young-in-shape bras are pretty with embroidery, practical in cotton and easy on an allowance. Could be, you'll take all three!

at leading U.S. and Canadian stores and shops or write us, we'll tell you where

left to right, all in white:
#1206 comes in AA cup, $1.75
#1313 strapless bra molds gently with foam rubber, holds securely with light boning. AAA and AA cups. $2.50
#1312 is lightly padded. AAA and AA cups. $2.50

GOSSARD

Gossard quality line of Pixie

THE H. W. GOSSARD CO., 111 North Canal Street, Chicago 6 • New York • San Francisco • Atlanta • Dallas • Toronto

6
SHAMPOO, MACARONI, AND THE AMERICAN GIRL: SPRING COMES EARLY

> Still she went on growing, and, as a last resource, she put one arm out of the window, and one foot up the chimney, and said to herself, "Now I can do no more, whatever happens. What will become of me?"
>
> — LEWIS CARROLL,
> *Alice's Adventures in Wonderland*

On the afternoon Sonya Lunder came to visit, I was, as usual, madly making after-school snacks for the kids, who are seven and nine. It entailed scooping some dip out of a store-bought plastic container, peeling carrots, and slicing some cheese. The kitchen is where love and guilt — those dual sirens of parenthood — merge at their most intense. It was a fitting place to begin planning a chemical-body-burden experiment. At the heart of the experiment lies one central fact: girls today are getting breasts earlier than ever before.

Lunder works as a senior toxins analyst for the Environmental Working Group. The mother of two small children, she has taken her knowledge of chemicals into realms I never imagined. At the gym where our kids take classes, she noticed the big foam blocks in the "pit" that students land in as they jump off a rope swing. She asked the gymnastics director if the foam was treated with flame-retardant chemicals, which are known to slough off and accumulate in human tissues. To his credit, he'd actually considered the question before and told her he tries to get untreated foam. She encouraged him in her friendly Sonya way to keep trying.

In my kitchen, Lunder zeroed right in on the fridge. Even the door's water dispenser attracted her attention. "You know, the water tubing is probably plastic," she said. After noting all the plastic-contained food in my fridge — yogurt, juice, cheese, meats — she pointed to a cutting board on the counter. She asked if the original packaging had said "antimicrobial" on it. I shrugged. "If it did, that means it's embedded with triclosan," she said. She scrutinized my hand soap gel. I thought she'd be happy, because it was from a leading "green" brand and declared on the bottle that it contained no parabens. "Interesting!" she said, reading the smaller print. "It's got benzophenone in it! That's related to a sunscreen chemical linked to endocrine disruption in animals. Why would they put that in your hand soap?" she mused. "It probably keeps the contents from breaking down in UV light."

Lunder was here to help me prepare for the detox phase of my home experiment, in which I would valiantly try to avoid exposures to chemicals to see if I could reduce my body's levels of toxins. The experiment, designed with help from both Lunder and Ruthann Rudel, the toxicologist for the Silent Spring Institute, would incorporate a before and an after phase for both me and my seven-year-old daughter, Annabel. We would compare our levels to those of five other American families in a study sponsored by Silent Spring and the Breast Cancer Fund, both nonprofits interested in elucidating the environmental connections to breast cancer. In the first "tox" phase, Annabel and I would live our normal lives, absorbing hidden molecules of plastic from our food and the occasional canned beverage, as well as from shampoo and moisturizers purchased from the supermarket. In other words, for three days, we would be average Americans. Then we would pee into some glass jars and ship them to Canada for testing. A week later, I would enter detox for three days. I would become a vegan, avoid triclosan and plasticizers to the best of my ability, and have my urine retested. I would not drive (to avoid chemicals in the car upholstery and interior) and would only eat food that had never touched plastic. As a testament to just how foreign this is to the American experience, I couldn't imagine getting Annabel to undergo full detox. It would mean no vehicles, no pizza, no bubble bath. I didn't have the heart to put her through it. For much of phase two, I'd be on my own.

Why bother with all this? Blame it on the government. For seven years, the National Institute of Environmental Health Sciences (NIEHS) and the National Cancer Institute have joined forces in a $40 million study trying to understand the forces behind early puberty and its link to breast cancer. I'd heard about the effort, called the Breast Cancer and Environment Research Centers (BCERC), and was closely following the results, which fall into three main buckets: chemical exposures, lifestyle (diet and exercise), and social factors (economic status, number of parents in the home, etc.). Data have been coming out of four research hubs across the country, in which a total of 1,500 girls were participating, beginning at the age of six. Already the information shows these girls have some of the highest body burdens of everyday chemicals yet seen. As a group, they also have the youngest breasts ever recorded. I was curious to know how our levels matched up, and what it all meant.

Scientists have known about the relationship between the age of puberty and breast cancer for a long time, ever since they began looking at lifetime estrogen exposures. Then in 1997, Marcia Herman-Giddens, a scientist at the University of North Carolina, published a bombshell of a paper. Her research indicated that girls were developing breasts and sprouting pubic hair one to two years younger than expected, with white girls getting breasts at a mean age of 9.8 years and black girls at 8.8 years (menstruation typically starts two to three years later). Within two years, experts revised the official definition of "precocious puberty" downward, from age eight to seven in white girls and from age seven to six in black girls. Welcome to the new normal.

The age of puberty matters. A 2007 report

for the Breast Cancer Fund notes that if puberty can be delayed by one year in girls, thousands of breast cancers could be prevented. Suzanne Fenton, a scientist with the Reproductive Endocrinology Group at NIEHS, went even further. "We think that puberty may be the main driver in the risk of breast cancer," she told me. If you get your first period before age twelve, your risk of breast cancer is 50 percent higher than if you get it at age sixteen.

Why? It might have to do with hormones. Earlier puberty means an additional few years of estrogen and progesterone flooding the breasts and causing cellular changes. Or it could be that estrogen gets metabolized by the body into some toxic by-products that create free oxygen electrons and can damage DNA. But why would an extra year or two have such a big effect given our long lifetime of ovulating? Another theory is that puberty — with all its inherent cellular instability — causes breasts to be extra vulnerable to carcinogens. If puberty comes earlier, that window of vulnerability is open for longer. The new government-funded research attempts to answer some basic questions: What causes early puberty? Is it something we can possibly manipulate?

I got my first period when I was eleven.

Annabel is likely to fall on the early side too, since the magic number is driven partly by genes. At seven, she is the perfect age to examine suspected environmental factors. We're testing for some of the same chemical candidates investigated in the BCERC study: BPA, phthalates, triclosan, parabens, and flame-retardants, all common substances known to mimic sex hormones or otherwise interfere with their normal coursings through the body.

When I look at Annabel, I see total pre-pubescence. She sings and twirls around and makes elaborate rooms and nests for her dolls and stuffed animals. She is still drawing figures that are not anatomically endowed. It's almost inconceivable to imagine that some girls her age are actually wearing bras. To put the data into perspective: by 2011, one-third of black girls between the ages of six and eight were "budding" breasts (that's actually the technical term) or growing pubic hair, along with 15 percent of Hispanic girls, 10 percent of white girls, and 4 percent of Asian girls. Some of the breast budding is correlated with obesity. We live in the thinnest town in America — Boulder, Colorado — and not many of the mostly white, professional-class kids in Annabel's elementary school are overweight.

Still, the fourth- and fifth-graders are maturing here earlier than they did when I was their age. Today, half of all girls in the United States start popping breasts by their tenth birthday, well ahead of girls of the *Brady Bunch* era.

If you're a parent, this should make you nervous. Puberty has got to be one of the strangest and most profound experiences we humans undergo. Breasts and hair appear out of nowhere, voices drop, personalities change. We get hijacked by hormones and taken to another planet. It would all be comical if it didn't involve teenagers doing so many stupid and hazardous things involving their bodies, high-speed vehicles, and illicit substances, sometimes all at the same time. Girls' bodies begin to enter an uneasy and public arena of male attention. We probably all remember that exquisite tension between wanting to show off our new breasts and not wanting anyone creepy to notice them.

I started getting breasts around the summer between fifth and sixth grade. That's when my father gave me a book about girl power. His girlfriend, a librarian, had recommended it. The gist of it was how to say no to boys or men who wanted you to do things you didn't want to do. I was uncomfortable reading it. I was uncomfortable discussing it with him. I just wanted to think about my new blue Princess phone and calling my girlfriends.

By the next summer, when I was twelve, I had small but shapely little mounds. I was now tall enough to wear some of my mother's clothes, a rite of passage in itself. She had a designer T-shirt, an incredibly soft, thin weave in orange and green stripes. It must have shrunk in the wash because I don't think it could have fit both of us. She was a 36D, and I was probably a 28AA. But the shirt fit me, and I claimed it. It was a treasure. Between the thin cotton and the stripes, I knew it made the most of what I had.

That summer I traveled out West with my father and some family friends, including two teenaged boys. I wore the shirt often. One day we hired a guide and some horses and rode out to see some ancient ruins in the Arizona desert. It was a long ride over rough country. On a couple of occasions my horse lagged behind and out of sight of the others. Our guide, a youngish Navajo cowboy, would come galloping back to me, reach out, and give my nearest breast a squeeze. He would roar with laughter and try to do it again. I rued my slow horse, my striped shirt, and my posse for leaving me behind. I goaded that old horse until I finally caught up with the others.

A few months later, in the completely different canyons of New York City, I took a taxi with two friends across town. After we paid our fare, the driver reached back through the open Plexiglas window and groped us as we slid across the seat to the door. That time I cursed him out and slammed the door. Maybe my father's assertiveness training had gotten through. I was now in the brave new world of boobs.

How many girls have memories like this, or worse? Probably all of us.

It's hard enough to stand your ground as a preteen. But an eight-year-old is pretty much incapable of it. Little girls aren't emotionally equipped to handle the challenges of puberty. Teens barely are. Girls with breasts are the targets of teasing and jokes by their peers, and statistically they're more likely to be victims of sexual assault. Studies have shown that prematurely developed girls are at greater risk of substance abuse, depression, and suicide.

Lately, the advent of earlier puberty has converged with the advent of 3G and 4G video-phone technology. The tragic result, according to pediatrician Sharon Cooper of the University of North Carolina School of Medicine: more cyber-sexual exploitation of very young girls.

While the declining age of sexual maturity is worrisome, it's important to know that it's not entirely new. Here's the general evolutionary scenario: if you can eat more, you can reproduce earlier, resulting in more offspring. And in fact, our ascent out of caves and hovels has meant that the age of sexual maturity in girls has dropped slowly but steadily, about three months per decade, since 1850. The reasons for the change were expected and benevolent: better nutrition and less infectious disease. But does this mean that we're designed by nature to have babies as early as we possibly can? Not necessarily.

Humans, as we've seen, are unusual creatures. One of the ways in which we are unique lies in our prolonged childhood. We take longer to reach sexual maturity than any other primate, meaning we sit around for more time being relatively useless with regard to the propagation of our species. (Chimps, by contrast, become mothers when they're six to nine years old.) There are good reasons for our dawdling; namely, our bodies and brains are big and grow slowly. The longer we hold off puberty, the taller and stronger we'll be and the greater the likelihood our offspring will be healthy too. (One of the reasons boys grow taller than girls is that they reach puberty later. The sex hormones that come with puberty, such as estrogen in girls, actually seal the bones from growing more. Sometimes estrogen is given to prepubertal girls by doctors who believe they are growing too tall too soon, which will likely go down as a *bad idea* in the annals of pediatric medicine.)

Having a long childhood, though, takes a toll on parents and society. Kids need resources, a whopping 13 million calories approximately, before reaching adulthood. For some populations of humans — those with fewer resources — that toll is harder than for others. This may partly explain the great natural variation in age of puberty from eleven or twelve to sixteen. Although girls tend to reach puberty about the same time their mothers did, there's still a lot of wiggle room.

Here's where we give another nod to the power of our environment. Breasts are like little antennae, processing information out there in the neighborhood and bringing it home. Among other things, the environment tells breasts when to show up. Evolution designed our bodies to be responsive to all sorts of cues, including the nutritional status of our mothers and even, remarkably, our grandmothers.

Even though genes account for some of the timing of puberty, the environment determines when and how the genes get switched on. When a woman's prenatal nutrition is good, her baby's fetal cells program themselves for earlier puberty. When it's lousy, the opposite occurs. And along the way, there's still opportunity for course corrections. When poor, once-hungry immigrants move to industrial countries, suddenly encountering "nutritional excess," the girls' bodies recalibrate. Adopted Indian girls who move to Sweden as infants, for example, reach puberty uncommonly early. They also gain weight faster than their Scandinavian peers, having been programmed in the womb to store more calories as fat.

Okay, so now we know the age of puberty naturally fluctuates. That doesn't mean we should shrug it off. The Herman-Giddens data suggest that between the 1970s and the 1990s, the age of puberty dropped much faster than expected. Perhaps the latest oscillation isn't so "natural" after all.

There's no doubt modern society, with its

overnutrition and better medicine (as well as perhaps its industrial pollutants), has caused the age of sexual maturation to fall. Ironically, though, this same complex modern world means we need *more* time for what experts call "psycho-social maturation," or time to figure things out. The human brain takes two decades to mature, especially the prefrontal cortex, which governs self-concept and the control of impulses and reward-seeking. It's arguably a lot harder to be a young mother now — with all our social and familial fragmentation — than ever before. Evolutionary biologists call this a "mismatch." Our bodies are responding to one set of determinants, but our brains don't appear to be catching up.

Our Paleolithic legacy has left us with other mismatches, such as obesity and diabetes. It's becoming increasingly obvious that our bodies weren't designed to handle modern, industrial diets. While metabolic diseases have received a lot of attention, the changing age of puberty hasn't. But it should. Peter Gluckman is a biologist at the University of Auckland in New Zealand and a pioneer in a field known as evolutionary medicine. As he puts it, "For the first time in our evolutionary history, biological puberty in females significantly precedes, rather than being matched to, the age of successful functioning as an adult."

What exactly is it about our modern environment that's driving the age of puberty toward toddlerhood at an ever-increasing pace? The government's BCERC study posits the broad hypothesis that "chemical, physical and social factors interact with genes" to set breasts on their inflationary course. Many of those factors are new in our human life history. The research is divided into three areas: animal studies, human breast cell studies, and real-life, whole-person studies. It may be our best hope for understanding some of the key unanswered questions behind the mysteries of puberty and breasts.

Although it's tempting — for both the researchers and people like me — to try to identify a single culprit for early ballooning bosoms, we won't be easily gratified. One thing seems clear: many factors lead to the complex, multistaged "event" of puberty.

Obesity has long been the leading contender in the early-puberty sweepstakes. Everyone in the field knows that the percentage of American girls aged six to eleven who are obese more than tripled, to 16 percent, between 1974 and 2006. Nearly a third of girls are now labeled overweight. Although the exact timing of puberty remains wrapped in mystery, the role of fat is partly known. Where there's fat, there's also the enzyme aromatase. Aromatase helps convert the building blocks of steroids — cholesterol — into estrogen. Fat has been called "the third ovary" because of its ability to make estrogen. The crude equation is more fat equals more estrogen, and this is true even in men, witness man boobs. Fat also increases levels of leptin, a hormone that tells us when we're hungry, and also seems to fire up the puberty engine. Leptin levels have been found to be higher in African Americans than in other groups. But fat is not the whole story, not by a long shot. Otherwise, fat babies would be oozing in estrogen, and they're not. And obese women often have *less* circulating sex hormones because they don't ovulate regularly.

Puberty is like an orchestra. The ovaries are the violins. Fat cells are the oboes. We have some sort of internal conductor that tells the parts when to make their music. Some call it a gonadostat — literally, a thermostat for our sex gonads — and it sets our pubertal tempo. Whatever it is, it's regulated by the hypothalamus in the brain, the director of the symphony, and that in turn responds to all sorts of internal and external cues in the forms of enzymes and hormones.

The BCERC researchers are devoting a lot of time to studying diet and exercise, regularly asking the 1,500 girls what they're eating and how often they visit playgrounds, play sports, and walk to school. So far, there appears to be some relationship between the amount of fiber in their diet and the age at which they get breasts. The more fiber and vegetables they eat, the later they enter puberty. Another study out of Britain in 2010 found that girls who reached puberty earlier ate more meat than their peers, with the biggest carnivores maturing earliest.

Said Dr. Frank Biro, a pediatrician and coinvestigator of the BCERC study from the University of Cincinnati, "The nutritional factor consistently associated with timing of puberty, in those societies where there are sufficient calories for all members, is fiber. Higher fiber equals later maturation." This is probably because of the way some genes interact with fat, he says.

Fair enough, but as Biro himself acknowledges, diet alone doesn't seem to explain the puberty clock. I never totally bought the simplistic fat-triggers-puberty argument. I went through puberty a good year earlier than all of my friends, and I was skinny as a rail and always had been. Doctors and cancer scientists know that plenty of thin girls fall on the early side of the puberty curve.

It turns out that while fat may trigger one pathway to puberty, there are other pathways as well. Some girls get breast buds as their first sign of puberty. It may be months or years before they grow pubic hair or get their periods. Other girls grow pubic hair first, with nary a breast in sight. Pubic hair is influenced more by adrenal hormones than by estrogens. So you might think this means you're off the hook for breast cancer if you were thin, but, maddeningly, it doesn't. In fact, thin girls who menstruate early are at slightly higher risk of breast cancer later on than their pudgy peers. In a recent Swedish study, researchers found that women who had fatter bodies during childhood were 27 percent *less* likely to have breast cancer than women who were leaner as children. I was tall and thin as a fourteen-year-old, and it turns out both of those traits are linked to higher breast cancer risk.

Because of the holes in the obesity theory, I was curious to visit the lab of pediatric endocrinologist Lise Aksglaede in Denmark. Her office at the Rigshospitalet of the University of Copenhagen looks out over the west side of the city and its lakes. She was late for our appointment and came bounding in, breathless, from her hurried bike ride across town. Like nearly one-third of all workers in the city, she bikes to the office every day. And like nearly everyone here, she's fit, blonde, and peppy. I suspect the fit and peppy parts come from all the biking. According to various happiness indices, the Danes are among the happiest people on earth, despite terrible North Sea weather and a crippling tax code. In her mid-thirties, Aksglaede is still breast-feeding her one-year-old son, chalking her up as another statistical norm; breast-feeding rates in Denmark are double those in the United States.

Wondering whether the American puberty data were really so unusual, she decided to take a closer look at her hometown. She and her colleagues examined almost one thousand girls in 2006 using the exact same protocols as a similar study in 1991. It turned out that the girls — all white and middle class — started budding breasts a full year earlier than they had just fifteen years ago. (The age of menstruation had advanced only about four months.) But the real head-scratcher was that the change in the girls' body weight was minimal and couldn't account for the difference. Nearly all the girls were relatively thin, said Aksglaede.

So if fat isn't setting the puberty clock for these girls, what is?

There are three other leading theories: artificial light, divorce, and highly sexualized media. The light theory hinges on melatonin, a hormone that flows from our pineal gland at the center of our brain to our hypothalamus, telling it to quiet down (the hypothalamus, remember, regulates our gonadostat). We make melatonin in darkness. Women who are blind make more of it, and guess what? They have a lower risk of breast cancer. By contrast, women who work under lights on the night shift at work have a higher risk.

Nighttime light — in the form of computer and TV screens, electric overhead lights, and the ubiquitous hallway night-light — is clearly not something we evolved with. Some researchers speculate these girlhood light exposures could be suppressing natural melatonin levels, and in turn speeding up the gonadostat. Studies have found that girls with precocious puberty have unusually low levels of melatonin, while female athletes have high levels. (Note to self: remove the nightlight from Annabel's bedroom.) The problem, though, with the light theory is it doesn't totally explain the difference in just fifteen years in girls in Copenhagen. Presumably, even with their crippling tax code, Danes had plenty of electric lights in 1991.

Let's look at the divorce theory: it's been documented that girls not living with a biological father tend to mature earlier. Female elephants do something similar in the absence of a parent. It makes some survival sense that populations under stress would be more desperate to reproduce, even if they can't raise the next generation in an optimal environment. The absent-father theory could account for my early age at puberty. My parents divorced when I was two. But it couldn't explain all of today's girls, because families are actually slightly more stable now than they were two decades ago.

Which leaves us with the boobs-and-sex-all-over-the-media theory. It sounds reasonable, but if girls are becoming more sexualized from a constant stream of media images, their hormone levels would also be rising, and oddly, that's not happening. While the Copenhagen girls are growing breasts earlier, their bodies are not making any more estrogen than they were in 1991. To Aksglaede, this indicates that the source of estrogen — needed for breast development — must lie somewhere else.

"Our best suggestion is that [the source] is something from outside," said Aksglaede. "The main discussion is environmental factors." Specifically, chemicals that mimic hormones, many of which girls are exposed to every day, even in Europe, which is only now just starting to regulate them. Endocrine-disrupting compounds, as we saw in the last chapter, include the much-publicized baby-bottle ingredient BPA, as well as other ingredients in plastics, pesticides, and compounds in cigarettes, among many others.

The BCERC's Biro agrees, to a point. "Heavier girls are more likely to enter puberty first," he said, "but something above and beyond that is going on and that's where it gets really interesting. There are lots of others who believe that chemicals are the major cause. I believe that they are clearly contributing."

Not all experts are convinced. Dr. Paul Kaplowitz is the man behind revising the clinical age of "precocious puberty" after Herman-Giddens's groundbreaking study came out in the 1990s. "The environmental puberty hypothesis is interesting," said Kaplowitz, the chief of endocrinology and diabetes at Children's National Medical Center in Washington, D.C. "But my position is that we need more information on environmental exposures. I've gotten into the habit of asking my patients with early puberty if they're using hair care products, essential oils, lavender, tea tree oils, and so on. It's pretty rare for them to say yes. The phthalates and the BPA are plausible. But if external estrogens were really affecting girls, you'd think you'd see more breast development in boys as well. We do see some of that, but I haven't seen a big increase," he said.

Biro counters this argument by pointing to studies suggesting that boys are indeed showing signs of unusual estrogen and anti-androgen exposure, such as smaller penis sizes, decreased sperm counts, and shorter distances between the genitals and anus, all of which are considered markers of "feminization." By some estimates, the once-rare birth defect of undescended testicles in baby boys is increasing in the United States and parts of Europe. In a study of 1,600 babies born between 1997 and 2001, Danes had smaller testicles than the Finns. Scientists know this because they expertly measured "ellipsoidal volume" and found the Danish package lagging at birth. The differences were even more pronounced after three months, with the Finns averaging *three times* more testicular growth. Researchers went back and tested samples from the babies' stored blood and their mothers' breast milk in each country. Danes are known to smoke and drink during pregnancy, but that didn't seem to explain the genital effects. Then other hormone-monkeying suspects turned up at relevant levels: certain industrial chemicals. As researcher Katharina Main, who works across the hall from Lise Aksglaede at the University of Copenhagen, told me, "It's higher here. The higher your [chemical] burden . . . the higher the risk of undescended testes."

It's been well documented that puberty can be swayed by chemical exposures. Lead and dioxin (a by-product of combustion), for example, are known to delay puberty in both animals and people. The pesticide DDT has been associated with earlier puberty in girls, and PCBs (polychlorinated biphenyls, used as industrial greasers) have been alternately linked in separate studies with both delayed and advanced puberty.

But external culprits are hard to pinpoint. Consider the weird epidemic of precocious puberty in Puerto Rico during the 1980s and 1990s. So many very young girls (under age five) were developing breasts that the government there established the world's only "early sexual development registry" to record it. The incidence there turned out to be eight out of every thousand girls between 1984 and 1993, eighteen and a half times higher than the incidence in the United States. Why would Puerto Rico be the world's hot spot for toddler breasts? The cause appeared more geographic than genetic, because Puerto Ricans in the mainland United States had normal development, and the problem was afflicting other ethnic groups living in Puerto Rico as well.

Journalist Orville Schell describes interviewing a Puerto Rican pediatrician and seeing Polaroids of some of these children in 1982:

In the first photo, a four-and-a-half-year-old girl with delicate coffee-colored skin, doelike brown eyes and almost fully developed breasts lies on an examining table. She smiles with a sweet innocence at the camera, seemingly unaware of the dramatic changes that have gone on in her body.

"She had an ovarian cyst," says Dr. Saenz [de Rodriguez] tersely.

A twelve-year-old boy stands against a white wall looking with blank bewilderment into the camera. He wears a silver crucifix around his neck, which dangles down between two grossly swollen breasts.

"We've had to schedule him for surgery," says Dr. Saenz matter-of-factly. "The emotional stress on him is incredible."

The children had reportedly eaten school-lunch chicken contaminated with high levels of DES, the same hormone Pat Hunt's mother took to prevent miscarriage, as we learned in the last chapter. At the time, DES was given to farm animals to make them plumper. When many of Dr. Saenz's patients stopped eating meat and milk, their symptoms improved. But an investigation — conducted after a considerable delay — turned up only average levels of the growth hormone in more than eight hundred subsequent food samples. Investigators also considered pharmaceutical wastes and agricultural pesticides, both widespread there, but again, there was no obvious evidence. One study did find unusually high levels of phthalates in the blood of girls with early breasts, but there is the lingering possibility that those samples were accidentally cross-contaminated by plastics in the testing laboratory.

Phthalates are a family of molecules used as scent stabilizers in lotions and shampoos, and as common additives in plastics. Concerned over their health effects, the European Union will soon be restricting three types of phthalates from general use in commercial products. California recently banned several phthalates in products sold to the under-three set. Known endocrine disruptors, they've been recently linked to genital abnormalities in baby boys and increased "girl play" in toddler boys. It's possible they may be turbocharging the future breast tissue of young or even unborn girls. One way to find out is to compare the urinary levels of phthalates in girls and see who gets the earliest breasts. (Unfortunately, it's not enough to simply ask the girls which toys and personal care products they've used; companies are not required to list ingredients on their labels. The only reliable way to test exposures is to ask the girls for body fluids.)

The BCERC researchers are measuring

girls' blood and urine for fifty-one chemical substances, including phthalates, BPA, organochlorine pesticides, heavy metals, industrial solvents such as PCBs, and naturally occurring plant-based estrogens such as soy. Nearly all the girls have these substances in their bodies, sometimes in much larger concentrations than what is found in adults. There are also some geographical ticks: Girls in California are endowed with the highest recorded levels of flame-retardants, probably due to that state's strict flammability standards. Girls in New York City carry more cotinine, a molecule found in cigarettes and second-hand smoke, as well as more 2,5-dichlorophenol, used in mothballs and room deodorizers, while girls in Cincinnati carry more PFOA, a chemical used in making products with Teflon.

The Centers for Disease Control and Prevention (CDC) releases measures of the so-called body burdens of girls every two years as part of its National Health and Nutrition Examination Survey. When the BCERC researchers in Cincinnati studied the values measured in their girls, they found that fully 40 percent of them carried levels of PFOA above the CDC's last measured ninety-fifth percentile. The researchers are currently trying to determine why the girls had unusually high levels and whether PFOA is linked to early breast development, particularly in thinner girls.

African-American girls in the study bore considerably higher levels of phthalates (the fragrance stabilizer) and parabens (a preservative in personal care products), four times the level found in white and Asian girls. Could this possibly help explain why the proportion of African-American girls who reach puberty at age eight is four times greater than the proportion of white girls? Or why African-American women in their twenties have a nearly 50 percent higher rate of breast cancer than white women of the same age (above age forty, though, the white rate surpasses the African-American rate). Researchers would like to know.

As I found out from our home experiment, phthalates are everywhere. Since they bind fragrances, they're often present in our shampoos, soaps, and moisturizers. They are an important ingredient in softening polyvinylchloride (PVC), so they waft out of things like shower curtains, plastic toys, and fake leather. (There's more fake leather lying around than you think. I had to throw out one of my daughter's play wallets because it smelled so strongly of chemicals. My son has a belt made out of it, and it's also common in sneakers. Now I have a radar for "pleather" when I go into box retail stores, and I steer clear.) Phthalates may also be present in sandwich bags and plastic wrap, but it's hard to know for sure since manufacturers aren't saying.

A newspaper in Taiwan recently reported that phthalates had been found in food such as baked goods. Apparently it makes them smell good even after many days on the shelf. This put officials in the absurd position of warning consumers not to buy food that smelled fresh.

For the before — or tox — phase of our experiment, I indulged in some very American habits, and Annabel joined me in some of them. For three days, we ate a couple of meals out of cans. (One out of every five U.S. dinners includes a can, according to the Canned Food Alliance.) Refried beans were a favorite. I drank a can of ginger ale at lunch and dinner. I indulged in a professional pedicure (pearly pink), the better for sitting briefly in a cloud of chemicals. We both shampooed and conditioned with "fresh" floral fragrances and used a perfume-y bar of soap. I wore brand-name deodorant, moisturized my body in "deep healing" lotion. I topped it off with a swath of coral lipstick. I felt like a beauty contestant.

As anyone on *Celebrity Rehab* can attest, detox isn't nearly as fun as what precedes it. It was very hard to avoid food that had never touched plastic. Coffee beans? Nope. Grapes? Sorry. I rode my bike (because my car interior off-gasses phthalates as well as flame-retardants) to the farmers' market. Even my bike had plastic handle grips, but there was no helping that. I also wore a helmet made of shatter-proof polycarbonate, rife with BPA. (Yes, there are plenty of beneficial uses for BPA.) My prescription sunglasses are made out of the same, another product that was hard to avoid considering I left my metal frames behind in about 1999. At the market, I perused the farmers' offerings. Some, like the tomatoes, were displayed in plastic. No tomatoes for me. But beautiful artisanal breads were wrapped in paper bags. I asked the seller where the flour had come from, and if it had touched plastic. This being Boulder, he didn't look at me like I was a freak. He assured me the flour came straight from the mill in large cloth sacks, the old-fashioned way. I stocked up. I also bought some ridiculously expensive quinoa that came packaged in paper. I eyed some greens, but noticed the seller was serving them with rubber gloves (PVC) and stuffing them into plastic bags (phthalates). I asked him if I could handle my own greens and put them in my own cloth bag. He said sure. Boulder is apparently full of chemically sensitive plastophobes.

I had a good excuse not to do any other shopping: BPA is used to coat thermal paper, the kind used for store receipts and airline boarding passes. Unlike the BPA in polycarbonate water bottles, the BPA in paper isn't "bound" to other molecules, so it rubs off in relatively high amounts. Although we don't generally go around eating receipts, Sonya Lunder says the substance easily ends up on dollar bills and our hands.

By turning vegan for three days, I avoided meats and cheeses wrapped in plastic. I also knew I might drive my levels down by not eating animals that absorb chemicals from *their* food and water. The higher up the food chain, the more nasties you'll find. Large marine mammals are probably the most polluted creatures on earth. Regardless of how many chemicals I avoided, I lost three pounds.

So how did our body burdens measure up? We compared my levels to those of other adults in the Silent Spring Institute/Breast Cancer Fund's five-family study, as well as to the much larger CDC database. We also compared Annabel's levels to those found in the BCERC study of hundreds of other six- to eight-year-old girls.

For BPA, my before level, as measured in urine, was 5.10 nanograms/milliliter (ng/mL), or parts per billion. That level vaulted me just into the upper quarter of the American range (U.S. levels are, incidentally, twice those of Canada). My after-detox level was 0.80, a drop of 85 percent! Most BPA we consume only lasts about half a day in our bodies, so three days of plastic avoidance, fresh food, and veganism really worked. Annabel's levels went from 0.80 to 0.65, a drop of 18 percent. Annabel didn't have any soda during the before phase, which is perhaps why her levels started where mine ended. Interestingly, though, her levels dropped lower than mine during detox, maybe because of my polycarbonate glasses or some other mysterious exposure. Our detox BPA values beat those for most of the other families in the study, but we were unable to eliminate our exposure altogether. "You weren't able to go to zero, and that's consistent with other data out there," said Fred vom Saal, a biologist at the University of Missouri who specializes in BPA research. "We need to find out more about where this stuff is used."

My high before level is still considered plenty safe by the EPA. My level, in fact, fell about four hundred times lower than the agency's magic danger number. That should be reassuring, but vom Saal warned me against feeling smug. He and others argue that the EPA safe dose is woefully out of date, based only on a few thirty-year-old toxicity studies and not on more current research of "low-dose" effects on animal endocrine systems. In fact, vom Saal told me that my initial level of 5.10 is "getting toward the red zone in terms of being related to metabolic abnormalities [in animals]. Anything you can do to lower your exposures would be good," he intoned. Vom Saal's lab is currently investigating low-dose BPA exposure and urethra disorders. Guess I'll be rethinking those refried beans.

For triclosan, we got similarly spectacular detox action. The cutting-board chemical, triclosan is also added to soaps and other products as a disinfectant. Lunder's organization, Environmental Working Group, has sniffed it out in everything from toothpastes to deodorants to children's toys to shower curtains. Despite its dubious benefits, the "microbicide" has been a flat-out marketing success, appearing in 76 percent of commercial soaps. Unfortunately, it's also been shown to disrupt thyroid hormones in frogs and rats.

In humans, triclosan can be absorbed through the mouth and intestines, as well as through skin. It accumulates in fat and doesn't exit the body as fast as phthalates or BPA. The median level in U.S. children is 9.8 and in adult women, 12.0. In the BCERC group of girls, the median level was 7.2. Annabel's before level was considerably lower than average, 3.7, but mine came in at a whopping 141.0 after I amped up my daily hygiene routine with supermarket toothpaste, bubble bath, moisturizer, deodorant, and soap. Annabel joined me in the florid, floral bath. After detox a week later, she brought her levels down 48 percent, and I brought mine down 99 percent, to 1.3 (my underarms went *au naturel* and I used "natural" products on my teeth and skin). By tweaking a few of my habits over one week, I went from over ten times the mean to one-tenth of it. Still, why couldn't my ascetic ways zero it out? Probably because triclosan is now found in drinking water and food all around us.

Another notable chemical group we tested

was phthalates. There are many types of phthalates, and each has its own molecular weight and function in commercial products. Typically, the phthalates are measured by their metabolites, or what the molecules look like after circulating and exiting the body (again in urine). For example, one called MBZP, which is used to soften plastics, can be found off-gassing in car interiors. Another, MBP, is a breakdown metabolite of dibutyl phthalates, which, according to the CDC, are used in industrial solvents, adhesives, printing inks, pharmaceutical coatings, pesticides, and as additives in nail polish and cosmetics. Once these enter your body through your mouth or skin, half of them will be gone within 24 hours, at least until the next time you moisturize or eat. A memo released by the U.S. Consumer Products Safety Commission states that dibutyl phthalates "have become ubiquitous in the environment, and can now be found in food, water and air." Considered an "anti-androgen," MBP has been associated with genital abnormalities in rats and humans, decreased testosterone levels in men, and unusual mammary growth in male rats, among other problems.

The median daily level of MBP in U.S. adult females is 28.3 ng/mL. Hold your hat:

my before level was 375, or triple even the highest reported percentiles of all Americans measured by the CDC. But hold your hat one more time, and keep it off in tribute to one girl tested by the CDC: an eight-year-old Mexican American who logged in at staggering 101,000. Ruthann Rudel thinks this girl's parents should be notified, but that is not currently the policy of the CDC, especially when there is no clear clinical protocol for treating it, as there is for lead or mercury poisoning. "I want to help her, because I'm a mom and because this is a very high level," said Rudel.

Even after detox, my levels dropped to 63, still more than double the U.S. median. Rudel was stumped. "Both your levels were way higher than anyone else in the study," she said. Did I use nail polish remover right before detox? (I had.) Did I take medications? (Yes, for low thyroid.) Was I spending too much time with my printer? (Perhaps.) But Annabel's levels were also strangely high, finishing even above mine at 80 ng/mL, or double the median for girls her age as measured in the BCERC study. This is weird and, I'll admit, a little disconcerting. I have no idea why her levels were unusually high. Was she being exposed through food packaging in the school cafeteria? Was her toenail polish that potent? Is it because of the sunscreen I slathered on her for soccer practice? Then there was her school bus. While I was biking everywhere, she was sitting on vinyl seats.

Another phthalate metabolite, MEP, was also surprisingly high for me, 654 ng/mL versus the national median of 127. Annabel's was much lower at 18.2, but pity the young girl in the BCERC study whose level came in at 2,580. Preliminary data suggest MEP may be weakly linked to breast size in the girls. (But that could be due to related exposures in girls who wear a lot of fragrance, since many things in fragrances are estrogenic.) The girls with the highest levels had slightly more advanced breast development. After detox, I brought my level down only 66 percent. This metabolite comes from DEP, which is used in lotions, perfumes, and soaps.

Annabel and I also clocked high levels of other phthalate metabolites called MCPP and MEHHP. The former comes from vinyl gloves, garden hoses, cables, adhesives, and food packaging sealants. The latter can be found in products made with vinyl, including plastic wrap, toys, and consumer products. It has been associated with liver toxicity, decreased testicular weight, and testicular atrophy in rodents fed high doses. I was able to bring my levels for these down 62 to 95 percent, but Annabel only 5 percent, and her levels were higher than those in 95 percent of the girls tested.

For parabens, a whole other class of chemicals added as preservatives to cosmetics and to food products, I alone showed high levels followed by big drops. Females tend to have levels three to seven times higher than males. The CDC calls them weakly estrogenic and says human health effects are unknown.

What these tests tell us is how stunningly easy it is to get relatively high levels of biologically active chemicals into one's body. In just a few days of using mainstream toothpaste and deodorant, I scored off the charts for body burdens of triclosan and MBP. The good news is we know how to reduce some of our contaminants through better shopping; the bad news is it's so hard to do. What if you don't want to brush your teeth with an endocrine-disrupting pesticide? What if you'd rather not moisturize with printing inks and industrial solvents? Well, unless you have a chemistry lab in your basement, you're out of luck, because most labels won't tell you anything. (In the United States, foods, drugs, and cosmetics are exempted from federal reporting requirements.)

One lesson is that the "cleaner" you are — at least by the standards of consumer culture — the more contaminated you are. Another is that these are just a few of many, many biologically active compounds coursing through our bodies. "There are zillions of phthalates," said Rudel. "There might be some other important commercial ones that are endocrine disrupting. But they're not tested, so we don't know."

Partway into this project, I ran across a publication from the NIEHS. It described a talk given by George Bittner, a professor of neurobiology at the University of Texas, Austin, called "Are Plastics without Estrogen-Active Compounds Possible?" I found the title a stunner. It implies that nearly all of the plastics in our lives *are* estrogenic. And in fact, according to Bittner, that is the case, at least with common kitchen plastics. I gave him a call to learn more. Bittner and his colleagues chopped up hundreds of products ranging from plastic wrap to soda bottles and storage containers. Then they broke each of them down in a saline solution, and fed the extracts to estrogen-sensitive breast cancer cells. Over 90 percent of the extracts they tested made the cells grow, including many that were BPA-free. "The results were rather striking to us," said Bittner, who has founded a company to test plastics. "We had not anticipated it. There are hundreds, maybe thousands of chemicals used to make plastic that have estrogenic activity."

Of course, the question every parent wants answered is, What do all these exposures mean for our breasts and bodies?

Just because a chemical is sitting uninvited in your cells doesn't mean it's necessarily doing harm. This is a point the chemical industry loves to make: the amounts of chemicals released into our bloodstreams are so tiny, it's inconceivable that small quantities of additives from face cream or ATM receipts could be altering our bodies. The industry would prefer that people not even look for these chemicals; what's the point other than to cause needless anxiety?

These were the standard arguments used for decades to justify the unregulated presence of chemicals in the market. But scientists have two new tools with which to challenge the industry: first, they are using dazzling new technologies to measure small amounts of chemicals never before seen in our bodies; and second, with that advance comes the ability to see how these new-fangled molecules could be gumming up biological systems.

Tom Burke, the former director of science for the state of New Jersey, was one of the first people in the country to get his body fluids tested for the presence of industrial chemicals, and he defends the practice. Now the director of the Risk Sciences and Public Policy Institute at Johns Hopkins, Burke believes that the more people know about what's coursing through their bodies, the more likely industry will be to adjust the ingredients. "Sometimes the best management is a little sunshine," he said. Soon it will be as routine for people to test their bodies for these substances as it is to test their blood pressure. "We now know what different blood pressure levels mean and what diseases are associated with them," he said. "That is also where we are with lead and mercury and where we should be with other substances."

But we're not there yet. Just how might our moisturizers and sunscreens be causing earlier breasts? As we've seen, many of the molecules in everyday products look like estrogen, with structural rings held together by carbon atoms. It's possible that these substances are bypassing the body's normal hormone-making process, attaching directly to estrogen receptors in girls' breast tissues and switching them on before their time. Or they might be acting as "obesogens," altering gene expression that governs fat storage (making girls fatter, for example). Nobody yet knows how the molecules might be causing miscues in a growing girl's body. "What is scary is that we don't have any idea what the mechanism is. It's a big black box," said Aksglaede, the bicycling Danish endocrinologist.

To find out how these substances work in the body, the BCERC researchers are looking at their effects, both singly and when combined, in lab animals, from tissues to cells to genes. Seven years into the study, the lab science is complex and unsettling. For one thing, it's so new. It used to be that you took a sorry lab rat and dosed him with ever-increasing amounts of a chemical in question. When he keeled over and croaked, you knew you had a toxic effect. Soon scientists got better at looking for obvious signs of sickness, such as severe weight loss or tumors. Then they started being able to look at DNA, and to changes in DNA caused by toxins. Now, researchers aren't just looking at DNA mutations, but at how our environment triggers epigenetic changes

— how in "normal" DNA, genes can be turned off and on in ways that make lab animals behave differently, reproduce differently, parent differently, metabolize differently, and get sick in much sneakier fashion. They are noticing disturbing effects that would never have been seen in standard toxicity studies.

Some scientists argue that in addition to our genome, we should be mapping our "exposome," the environmental exposures that change our cell behavior. As we learn more about DNA expression, it's becoming increasingly apparent that human biological systems depend as much on external cues as on the code itself.

Drs. Jose and Irma Russo are conducting some of the most revealing experiments out of their Breast Cancer Research Laboratory at the Fox Chase Cancer Center in Philadelphia. The Russos met in a lab in Argentina and have hardly stepped outside one since, except to attend conferences, where they are coveted speakers. I first met them at a BCERC conference, and BCERC funds quite a bit of their work. "We wrote an abstract together in medical school," said Irma. "Our passion for science flourished into romance, and we've been talking about the same science ever since." Both in their sixties, they immigrated nearly forty years ago to work as pathologists at the renowned Michigan Cancer Foundation. They spent nineteen years there studying, among other things, the effects of estrogen on cell growth. Jose is small and trim, with short-cropped copper hair and glasses that seem to cover half his face. Irma is elegant and warm. Their maternal-nerd combo is perfect for nurturing a lab full of hard-driving international scientists and postdocs. It's a family affair. Their daughter worked as a tech there during college summers, and Irma makes sure the staff eats well during lunch meetings.

I visited them in their adjoining offices one snowy day in February. The oath of Hippocrates hangs on the wall between photos taken of their lab employees over the years and a few diplomas. Like good couples should, Irma proceeded to tell me about Jose's groundbreaking research, but she was frequently interrupted by Jose telling me about Irma's. For example, Irma pioneered using a rat model to study how the breast develops, because she found that rats have mammary glands (even if they have six of them) very similar to human mammary glands. Because breast cells are dividing like mad during puberty, the Russos have found that puberty is a very vulnerable time for these cells to be exposed to possible carcinogens. Their lab work has borne this out, over and over. Jose told me that "window of susceptibility" is a term Irma invented. "Around puberty, if something happens, there's an impact," said Jose in his thick accent. Added Irma, "If a girl has X-rays at twelve, she will have cellular damage."

I thought back to the X-rays I got at exactly that age for minor scoliosis. It seems somehow unfair that in addition to all the other troubles girls face during those tender adolescent years, even their cells are vulnerable.

Examples from the atomic bombs dropped on Hiroshima and Nagasaki are telling. Among girls who were exposed to the fallout, breast cancer rates (as determined decades later) were highest for the girls who were younger than ten when the bombs fell, and also high for girls between ten and twenty. They were lowest for women who were between twenty and forty. For many women, this would be after pregnancy has protected their breasts and before the increased vulnerability of menopause.

To further explore the unique windows of

harm for the mammary gland, the Russos have been dosing their lab animals at different stages of development, from prenatal exposure to prepubertal to later. One of the main chemicals they use to "assault" the rodents is the favorite egg-crusher of Patricia Hunt and one of our most ubiquitous daily-life substances: BPA.

In one particularly revealing experiment, they gave some young, "child-aged" female rats a hit of BPA, a bit higher than what we humans are exposed to everyday, and they left other rats alone. Then they let all the rats grow to early middle age, whereupon they exposed them to a known carcinogen called DMBA. The rats that had been given BPA before puberty grew more mammary tumors, and got them faster, than the control rats. To find out why, the Russos took apart the tumors and analyzed their genes. The BPA-dosed rats had altered DNA expression that supported cancer growth.

Here's a quick mini-lesson in cancer cell biology. For cancer to do its thing, a number of cellular events have to happen. These events generally fall into two categories: the *promotion* of cell growth (including gene transcription, replication, division, invasion, blood sucking, food gathering, and other delightful habits of a tumor) and the *suppression* of cell death (think of riot police who control unruly mobs, only now they're on strike). Genes in our bodies control these processes, and they are known, respectively, as oncogenes and tumor suppressor genes.

If we don't want cancer, we have to coddle our tumor suppressor police. We don't want them to go on strike, or get sick, or go rogue. This is, tragically, what happens to people who have the mutated breast cancer genes, BRCA1 and BRCA2. Those are technically suppressor genes, acting as DNA repairmen, and they haven't been working right in some families for thousands of years. One more recent variant of BRCA2 was traced back to a single "founder mutation" in Iceland in the mid-sixteenth century. It's a heck of a family legacy; around 80 percent of women born with bad BRCA genes will get breast cancer. The ones who don't get cancer probably have some other lucky genes to help them compensate, or maybe somehow they manage to skate through their developmental windows without any serious carcinogenic exposures. Women with the BRCA genes are much more likely to get cancer if they were born after 1940 than before, a fact attributed to everything from changing reproductive patterns, to body size, to the rise in the use of synthetic chemicals produced after World War II.

BPA appears to turn off some of our blessed suppressor genes and at the same time turn on the bad tumor promoters, the oncogenes. Some of us can resist these events better than others. Remember, the breast is the body's only organ that still has to do most of its basic construction well after birth. Therein lies some of our problem. With change comes instability. An organ incredibly responsive to cues both outside and inside our body, breasts are too trusting. That quality served them brilliantly in our evolutionary past, but it has not prepared them well for the modern age.

The Russos found that the rats exposed prepubertally were worse off than even the rats exposed in the womb, reinforcing the idea that the time near puberty, when the organ is growing fast, is the most vulnerable, at least to BPA. Their theory is that in order for breasts to form, there are "mammary progenitor cells" that are actively dividing at this time, and they are particularly responsive to hormonal signals. When their genes get messed up during this developmental tumult, they'll have a harder time warding off disaster later in life.

Jose's not-very-practical take-home message: "Avoid the exposure of young girls to these compounds."

If early puberty is, as writer Sandra Steingraber puts it, an "ecological disorder," what's a mother to do? Is it possible to keep Annabel on the childhood farm a little longer? Larry Kushi, a BCERC scientist and associate director of research at Kaiser Permanente, points to childhood diet as one area over which parents can exert some control. In recent experiments with adolescent rats, a high-fat diet led to inflammation, and later, cancer, in mammary glands. Kushi encourages more whole grains, more vegetables, and less meat. Kushi, the son of macrobiotic food entrepreneurs, is also a fan of tofu, which appears to have some preventative effects as far as early breast budding.

I have to admit, Annabel is not a huge fan of tofu. Or of whole wheat. She's part of the macaroni-and-cheese generation. I recently heard that children who watch "average" amounts of television will see five thousand commercials, more than half of which are about processed food. The "ad pyramid," dominated by fats, oils, and sweets, is exactly the opposite of the traditional government-recommended nutrition pyramid (now it's a nutrition plate). Forty-four percent of TV ads pitch sugary foods, while 2 percent pitch fruits or vegetables. It gives the term *boob-tube* a whole new meaning.

But some societal trends are pushing back. The public school system in Boulder recently stopped serving chicken nuggets. It replaced chocolate milk with organic low-fat milk. The cafeteria also switched to using organic, locally grown produce, all under the able tutelage of Ann Cooper, an Alice Waters protégée known as the "Renegade Lunch Lady." She started a Colorado-based nonprofit called Lunch Lessons to improve school lunches for the sake of things like health and attention spans. "We're done with Day-Glo orange cheese," she pronounced, earning my lifelong devotion.

Hoping Cooper would offer some other inspiration for my home cooking, I visited her at the school's district headquarters. Wearing a white chef's tunic, she ushered me into a cramped office filled with books. Was Cooper aware of the early puberty or breast cancer research? Not at all. But she's happy to add it to her arsenal of reasons to improve lunches for eight thousand public school children. It didn't surprise her that hormone-mimicking chemicals might be messing with our girls' bodies. Cooper, it turns out, is a DES daughter. She told me she'd been badly affected by the drug her mother took in pregnancy.

Under Cooper's direction, a gleaming salad bar now holds center stage in the school cafeteria. Back in the kitchen, the spaghetti and tortillas are whole wheat. The rice is brown. Hamburgers are served just once a month. I want to throw my arms around her in gratitude. Surprisingly, though, she said some parents are complaining about the loss of chocolate milk from the lunchroom. If Boulder Valley, with its culture of Buddhist triathletes, can't improve kids' diets, I don't know who can.

I've also changed some things in our own house. From our body burden experiments, I've learned to avoid scented products when possible. No air fresheners for us. No floral dryer sheets. No lavender shampoos. I don't want to make it sound like I'm a total killjoy, because that's not my style. We still use the occasional plastic straw and we still have plastic toys. I refuse to say to my children, *Step away from the Lego.* I still buy cheese in plastic because I'm not about to buy a cow.

In an ideal, childhood-preserving world, I would banish TV and avoid taking my kids to the grocery store with its eye-level barrage of snack temptations. So I'm working on that, but I'm not an all-or-nothing kind of person. I'm cutting down on canned beverages and foods. I could take a cue from Copenhagen's Aksglaede, who no longer washes her hair or uses sunscreen with products containing parabens or phthalates. (It's easier to be a savvy shopper in Denmark because that country requires labeling.) I now often pack my kids' lunches and snacks in thick glass containers and cloth sacks. This at times makes me feel virtuous, but more often than not, it feels like an exercise in futility. The food, after all, comes right out of a plastic carrot bag, bread bag, or cracker box. Even if I had the patience and fortitude to strap on a bonnet and grow the food myself, my local water supply carries a load of contraceptive and other hormonally active compounds.

In fact, the whole prospect of trying to individually safeguard one's family from silent endocrine disruptors feels like a folly, because it can't be done in any meaningful way until the government and chemical companies change the way they test, manufacture, and market these substances.

We can only eat so much quinoa out of a paper bag.

7
THE PREGNANCY PARADOX

The beginning is glorious . . . suddenly they begin to grow . . . breasts fantastic tender apricot breasts, then charming plucky firm tangerines, and then, just as you were on the verge of peaches, oranges, grapefruits, cantaloupes, God knows what other blue ribbon county-fair specimens, your stomach starts to grow and the other fruits are suddenly irrelevant because they're outdistanced by an honest-to-God Watermelon.

— NORA EPHRON,
Heartburn

Often, when a woman becomes pregnant, she first knows it by her breasts. They hurt and tingle as if someone just plugged them into a sound amp. And they grow, of course, doubling their weight and size, and the nipples enlarge and grow darker. I rejoiced in the physical changes of pregnancy. I loved my new growing shape, even liked the maternity clothes, except the granny underwear. I didn't gain much weight in pregnancy and it took me a while to start "showing." I was so thrilled when the first person who didn't know I was pregnant asked if I was. "I can just tell," she said knowingly.

I also enjoyed the bigger bustline, no doubt about it. Those new breasts cracked me up. It was as though I'd borrowed someone else's. I was partly right; the breasts weren't really the same. They weren't just my old breasts on hormones. Those county-fair specimens were in fact different breasts. It's something I've said before, but it's so cool I'll say it again: these are the only organ in the body that is not fully grown by adulthood. Breasts grow up when there's a baby cooking one story down.

Biologically, this is interesting. Medically, it's an epiphany. For at least several decades, researchers have known that the changes wrought by pregnancy protect those breasts from cancer. Lately, they've been wondering when we might be able to imitate, bottle, and patent that protection.

That quest has long motivated Irma and Jose Russo, whom we met in the last chapter, and Malcolm Pike, an epidemiologist at

University of Southern California and Memorial Sloan Kettering Cancer Center. They and others have been studying the protective effects of hormones on tumors and thinking of ways to mimic pregnancy, because, as Pike puts it, "making every girl have a baby at fifteen is just not going to happen." Perhaps the most we can ask for is a sort of medically induced hysterical pregnancy, or an immaculate conception minus the baby. If every young woman could just experience a simulated blast of pregnancy juice, she would be protected for life against breast cancer. Sure, she might have some strange cravings (and sore breasts) for a time, but well worth it. Pregnancy hormones, it turns out, are shockingly good cancer-preventing drugs, at least if you get the hit early in your reproductive life.

Here's what we know: a woman who has her first child before age twenty has about *half* the lifetime risk of breast cancer as a nonmother or a mother who waits until her thirties to have children.

These statistics have played out in my family as in so many others. My grandmothers and great-grandmothers were unusually early adapters to the modern notion of having children late. My mother's mother, Carolyn, wanted to be a judge. She graduated from Stanford Law School in 1926. She joked that there were so few women around, she could have a different date every night. Unlike most women of her generation, she was willing and able to put off marriage and children. After Stanford, she shared a law office in San Francisco with her stepfather. In her late twenties, she fell in love and married a young lawyer from Buffalo, New York. At twenty-eight, she had her first child, and then two more.

My other grandmother, Florence, a Chicagoan, married long after her compass was pointing firmly to spinsterhood. She fell in love late, with a tall Virginian, and had her first child at age thirty-three. These days that age seems perfectly normal, but in the 1930s the average age of first-time motherhood was 23.7. Before 1960, nearly one-third of American females had their first child before reaching age twenty. The pill changed all that. Since 1970, the percentage of women having their first children at age thirty-five or later has risen eightfold.

Carolyn, my lawyer grandmother, was first diagnosed with breast cancer in her sixties. She recovered, but the disease struck back in her eighties and ultimately killed her. Florence wasn't so lucky; just before her sixty-first birthday, she died of ovarian cancer, which is often linked genetically to breast cancer. Her mother had died of breast cancer in her fifties, and she had given birth to her first child at age twenty-nine. My mother, in contrast, had her first child, my brother, when she was eighteen, in 1950, well before effective birth control would become widely available to unmarried women. She didn't get breast or ovarian cancer, but she did succumb to a blood cancer in her sixties. Lately I've been wondering if her reproductive history might actually have helped her. Her early pregnancy interrupted her career, but it might have extended her life.

Knowing that I and so many of my peers had our first children well into our thirties, I'm both unsettled and intrigued by the power of pregnancy. The age at first pregnancy is now considered one of the most salient risk factors for breast cancer. In the life cycle of breasts, pregnancy is their defining get-out-of-jail card. If you draw it early, you've done as nature intended, and you're rewarded. If you arrive late or not at all, nature is less forgiving, for reasons scientists are now uncovering.

So what's so great about pregnancy, from a

breast's perspective? Aside from the happy boost in bra sizes, the answers aren't clear. The body is flooded with sky-high levels of estrogen, progesterone, and HCG, or human chorionic gonadotropin. This is what dime-store pregnancy tests measure, since HCG levels spike only in pregnancy. Different scientists disagree on exactly what the magic ingredient or combination of ingredients is and what exactly those hormones are doing. Something about pregnancy rebuilds the breast and armors it, by changing the architecture of either the cells or the proteins around them. One leading theory is that when the breast becomes fully mature by the end of pregnancy, its stem cells, which have been quiet, "differentiate" into cancer-resistant, high-performance dairy equipment. Even after weaning, the protection remains. But when the stem cells are waiting around for decades for their dance card, they're either weaker or just more likely to proliferate into cancer. And in a woman who's never been pregnant, her breasts remain undifferentiated for life.

"Women are so delicate," murmured Jose in his thick accent as he explained this to me.

"Yes, we are," shrugged Irma. "It's not sexist. It's our hormones."

In addition to studying the "window of vulnerability" that occurs around and after puberty, the Russos are also looking at the long cone of protection around pregnancy. Both pathologists, they are expert at picking out minute changes in tissue and cell structure and, in fact, seem to see things others can't. They first noticed dramatic changes occurring in rats, in which mammary cell changes are more obvious. Just as in people, lab rats that have been pregnant are less likely to get breast cancer. In one experiment, the Russos exposed "virgin" rats to a cancer-causing chemical. Some rats, however, had been primed with pregnancy-level HCG hormones. As the Russos suspected, those rats were the lucky ones. Their cancer cells divided less, their mammary cells shut down estrogen receptors, and they expressed more cancer-fighting genes.

I couldn't help but wonder if the pregnancy-as-prevention evidence might bolster the theory that many breast cancers are environmental in origin. If the breast is being "armored" in pregnancy, it's being armored against something. The lab experiments use chemical carcinogens to trigger cancer. The mother rats are protected while the virgin rats aren't. In humans, known genetic factors account for only about 10 percent of all breast cancers. On the one hand, you could argue that delayed childbirth "causes" cancer. But isn't that because something else is really causing cancer in unprotected tissue? I asked Irma about this.

She cautioned that we really don't know what causes breast cancer. But, she said, "from about the ages of twelve to thirty-five, from puberty to first pregnancy in many modern women, is a huge window to get exposed to radiation, alcohol, tobacco, phytoestrogens, xenoestrogens, and all the suspected carcinogens accumulating in the breast. If the same woman gets pregnant or gets the right hormones around puberty or a little later, the breast will mature and be more protected."

When I visited the Russos' lab in Philadelphia, their collaborator, Colombian scientist Johana Vanegas, showed me what was actually happening to the cells in the presence of pregnancy hormones. The rats' mammary glands had been removed and then put through a mini-meat-slicer-type contraption. Then the slices had been dyed dark pink and splayed out on small glass slides. I peered into an Olympus stereomicroscope to see some sections taken from rats that had never been pregnant. These "immature" glands looked like purple flower buds painted with a broad watercolor brush. Next I looked at the HCG-fed glands. These looked very different. The buds had transformed into the tiny petal dots of a pointillist painting. The immature glands represented by the watercolor image, at least in a rat, are called "terminal end buds," and they represent the undifferentiated state of cancer-vulnerable cells. Watercolor buds: bad. Pointillist petals: good.

The Russos say they can see the same thing happening with human breast sections under a microscope, although it's a controversial observation. They call the smooth never-pregnant glands "lobule type 1" and the late-pregnancy pointillist florets "lobule type 4." In the early and mid stages of pregnancy, you get the in-between lobule types 2 and 3. The higher the number, the more protected the cells. Among other things, lob 1 has more receptors for estrogen and progesterone, which, as we've seen, can be an entrance ramp on the cancer highway. Jose points out that the genomic signatures of the different lobes change dramatically as a woman cycles through pregnancy and beyond. Both full-on pregnancy and just a dose of HCG activate good-cop (tumor suppressor) genes, stop cell growth, and turn on other cancer-fighting genes, presumably for life.

Why for life? Because after pregnancy and lactation, the pointillist petals dissolve and revert to smoother buds, but these lobules tend to remain hardy types 2 and 3, say the Russos. If, as they speculate, cancer originates in lobule type 1, that would explain the protection offered by pregnancy: fewer type-1 lobules, fewer chances of cancer.

I'd had a gut sense that pregnancy was changing me in profound ways. Those mysterious hormones were bathing me in bliss, altering my core chemistry in a way that would prepare me for parenthood. Although I'd felt some ambivalence about what life would be like after having a baby, I'd wanted a child. I'd suffered a couple of years of miscarriages and failed conception, and now I was thirty-four. I'm sure part of my happiness was pure relief. But now I appreciate the power of the hormones, which seem, in retrospect, to have acted on nearly every cell in my body.

The thought of chemically mimicking this experience for a possible faraway health benefit gives me pause. But for some women, such as those who know they are carriers of breast cancer genes, playing with the ephemera of pregnancy might be worth it.

The Russos have begun testing their patented HCG regimen in eighteen high-risk women who have never been pregnant. The women will take HCG for three months, and then the Russos will compare their before-treatment and after-treatment genes to see if they've changed from a "high-risk signature" to a low-risk one. Natural HCG is made by the developing embryo, so the Russos get theirs from a synthetic process. HCG has some bizarre properties, including stimulating plant germination. The hormone has been tipping women off to their pregnancy status since 1530 BC, when Egyptians in the eighteenth dynasty watered seeds with their urine. If sacks of wheat and barley sprouted, a woman knew she must be pregnant. Today, some weight-loss clinics administer HCG because it's believed to help reduce abdominal fat when combined with a calorie-restricted diet (the better to feed a phantom placenta). MTV reality star JWoWW even sells it on the Internet. She gushes, "While on the HCG diet you will sleep sounder and feel better than you did before!"

In the Russos' trial, the women have noticed getting thicker, luxuriant hair, some weight loss, and a feeling of well-being and energy. "The ladies say they have a glow," said Irma.

Down at Texas Tech University Health Sciences Center in El Paso, pathologist Raj Lakshmanaswamy is a fan of using therapeutic estrogen rather than HCG to mimic pregnancy. (I know what you're thinking: if early puberty is any clue, estrogen is bad for breasts, but the reality is more complicated than this. While estrogenic substances may indeed be bad for the youthful, developing breast and the breast that already has tumors, estrogen has been given a worse rap than it deserves. I'll talk more about this in chapter 12.) Lakshmanaswamy envisions a hormone patch that women can apply for just three months in their early twenties to protect them from breast cancer. "We're talking about levels equivalent to the low end of pregnancy levels," he said. "It shouldn't cause any problems for that short a time. We're not there yet, but this is my feeling right now, that it can be done." Unlike the Russos, he hasn't patented this idea, saying he's more interested in basic science.

In LA, Malcolm Pike's team is looking at the breast tissue of women who have already received high doses of pregnancy hormones.

These women are patients in fertility clinics who are taking mega hormones to help them "hyper-ovulate," or produce a large number of eggs for in-vitro fertilization. "Does that change their breast?" asked Pike. "We don't know. You just have to do the hard thing, which is to study women. We do know egg donors have the breast-stimulating equivalent of three months of pregnancy in just one week of taking hormonal drugs. How does it happen and can you mimic it in smaller doses? They get a tremendous biological effect very, very fast. It's possible it happens in just two to three days. We need to check them again a year or two later and see if the same differentiation is there."

A major challenge in developing a fake-pregnancy drug is finding the right dose. "When you're pregnant, you have astro-nomical levels of steroids," Pike continued. "Absolutely astronomical. Before preg-nancy, you might have 100 units of estrogen in your blood, and when you're pregnant, you'd have 10,000 units or more. If I gave you that by mouth, you'd die. So a number of us are fiddling with it. It will still be a long time in the future. It's the early days yet of chemoprevention."

The pregnancy effect sounds like a slam

dunk: you get high levels of hormones, you're protected for life! Except that it's not a slam dunk. There's a lot of fine print. For example, the abortion exemption. You might think that if these pregnancy hormones are so great, then women who've had abortions are also protected, because they too enjoyed the spiking hormones of early pregnancy. The evidence, though, seems to suggest that this does not happen. Some years ago, a distinguished researcher named Janet Daling published results of a study suggesting that women who got abortions before the age of eighteen were *more* likely to get breast cancer, not less. A few other studies found similar results. The right wing seized upon this data, gleeful to have another reason to condemn abortion. Pro-life groups even sought legal action requiring that abortion be mentioned *as a cause of breast cancer* to any woman seeking abortion. Early in the George W. Bush administration, the federal National Cancer Institute's website proclaimed that abortion could increase a woman's risk of breast cancer.

Then in 2003, the National Cancer Institute convened a panel to sort through the evidence. It concluded that abortion did not increase a woman's risk, and that studies to the contrary were damaged by "recall bias," one of the notorious bad sisters of scientific method. Here's how these studies are typically conducted: You interview a bunch of women, say in their fifties, some of whom have had cancer and some of whom haven't. The catch is that the ones with cancer are much more likely to come clean about past indiscretions. In other words, as Pike described it to me, "abortion gives you breast cancer if you're Catholic, but doesn't if you're not." It appears the non-cancer Catholics simply lied about past abortions. Ah, the joys of epidemiology! No wonder these things are hard to sort out.

In any case, no one can claim that abortion *protects* you from cancer, nor do natural miscarriages. It seems a full-term pregnancy is needed for the breasts to fully differentiate. Which renders the high-dose, short-term faux-pregnancy therapies a big question mark, to say the least.

Around about the 1980s, some doctors began noticing an unexpected pattern: young women who had been pregnant in recent years were *getting* breast cancer. Rather than being protected by pregnancy, some women were experiencing the opposite. These women tended to be relatively older when they had their first child, and they tended to suffer from premenopausal breast cancer. What if the protective effect of pregnancy was just a myth or, at best, a historical relict?

In the mid-1990s, Pepper Schedin was, like so many other researchers, studying the storied protection offered by pregnancy. Everyone knew the breast goes through massive changes in pregnancy, but Schedin thought it might be worth looking at the massive changes that occur *after* pregnancy (or for those women who breastfeed their babies, after lactation), when the breast regresses back to a "resting" state. This process is called involution, or the massive loss of cells and structures that were part of the dairy machinery. In fact, 80 percent of the glorious pregnant breast gland simply disappears. Its ability to practically vanish overnight is yet another unique and strange feature of breasts. Schedin thought perhaps this was why mothers might not get breast cancer; perhaps nascent tumors were zapped out during this epic house cleaning.

She ran some experiments and found that while normal cells were indeed killed during involution, breast cancer cells were, startlingly, promoted. "Oh man, was that a surprise," she said. Just around that time, Schedin was contacted out of the blue by an old friend who had recently borne twins. The friend, who was in her thirties, had just been diagnosed with metastatic breast cancer. "I thought, huh, that's strange. It went against everything I'd ever heard. Pregnancy was supposed to be protective. Nobody ever mentioned it wasn't. So I went back and looked in the literature, and there it was: a small body of work on pregnancy-associated breast cancer, and no one knew why it was happening."

As far back as 1880, Samuel Gross, the surgeon subject of the celebrated Thomas Eakins painting *Gross Clinic,* noted that after pregnancy, breast cancer "was wonderfully rapid and its course excessively malignant."

The phone call changed Schedin's life. She now works in the young woman's breast cancer program at the University of Colorado's Anschutz Medical Center in Denver. Her office was decorated with framed photos of mammary gland cells and a giant poster of the well-known U.S. Postal Service breast cancer stamp. On a corner of her desk sat a 1915 microscope that her brother found in a junk shop.

Over the years she has made some interesting discoveries, most having to do with how the molecules of the breast talk to one another during involution. Remember, the breast gland doesn't just perch in an empty vacuum. It's a resident in a busy neighborhood filled with fat, collagen, and extracellular matrix, a rainstorm of proteins, hormones, and other material. Schedin has found that during involution, this matrix orchestrates a type of inflammation. Most of us are familiar with inflammation — it's what happens when a paper cut gets red and swollen or when we bump into the table and get a bruise. Immune cells rush to the injury and help repair it and battle infection. A similar thing happens to the retreating breast gland after lactation: macrophage immune cells swarm in to help clean the old gland and remodel the remaining tissue.

The problem is that sometimes our milk ducts have weird little not-quite-normal growths in them. Usually it's not a big deal, but sometimes these growths, or lesions, break free of the duct for reasons nobody entirely understands, tap into blood veins for nutrients and oxygen, and grow like bananas. Hello, cancer. This jailbreak appears able to happen during involution, promoted by the inflammatory environment. Schedin calls this the "involution hypothesis." It's just a theory, one of several, but she likes it. Older women are more likely to have these precancerous lesions in their ducts (perhaps thanks to their long years of environmental exposures); hence they're more likely to unleash cancer after their pregnancies.

So while young mothers may indeed be protected by pregnancy, old mothers are not. In fact, mothers who give birth after thirty have a slightly higher risk of breast cancer than women who never have children. That's right: if you heard nuns had it bad, older moms have it worse. And the types of cancers these moms get are more aggressive. A study in 2011 found that the more times a woman gives birth, the higher her risk of "triple negative" breast cancer. A cancer subtype making up about 10 to 20 percent of all breast cancers, these tumors do not express receptors for estrogen or progesterone, meaning they are more resistant to treatment and more deadly. (By contrast, postmenopausal breast cancers tend to be slower growing and can often be treated with hormonal therapies.) Women who have never given birth have a 40 percent lower risk of this type of breast cancer.

For the legion of us who had kids late in the game: bummer. Fortunately, pregnancy-associated breast cancer, called PABC, is still quite rare. In the United States, about 3,500 cases are reported per year, but under the standard definition a cancer has to be diagnosed within one year of pregnancy. Schedin fiercely disputes this definition and says pregnancy-related factors are still very much at work for many years after delivery. She thinks the risk goes up for five and maybe even ten years after pregnancy. "It's far more common than the stats let on," she said.

Sturdy and fit, with shoulder-length brown hair and glasses, she walked me through the eighth-floor lab overlooking east Denver. We passed a bank of freezers calibrated to −80 degrees Celsius (−112 Fahrenheit), the magic temperature for preserving the code of life, the RNA, in tissue samples. The tissue culture room smelled vaguely of cough syrup and sported a photo on the wall of a goofy-looking baby wearing a pink hat, below which exhort the words "Find a Cure before I Grow Boobs." The scientists here know they're working to help real people, thanks to their partnership with the university hospital and the young women (generally under forty) who proffer their cancer cells for research. In return, the lab tries to come up with therapies that will help these women before another forty years go by in the war against cancer. Schedin called this mission "Bench-to-Bedside."

If she's right and inflammation is causing trouble, Schedin wants to know what happens if you reduce it by taking ibuprofen, or fish oil, or other anti-inflammation substances. She's setting up a trial to find out. Another translation to the real world Schedin is willing to bet on: new mothers should get screened for breast cancer. Right now. They make up another high-risk group, she said, just like women over fifty or women with a family history of the disease.

She finds it unfortunate that the pregnancy-as-protective camp dominates much of the field. A street-tough Chicago girl who litters her words with expletives, Pepper is aptly named. "Not everyone agrees with me, but we need to let the science speak for itself," she said. "Pregnancy-associated breast cancer is too devastating to ignore." She's grateful that her work has led her to think of the breast in a whole new way, as a highly responsive organ whose signals get easily crossed. "I consider the gland plastic and poised to respond to signals because it needs to be quick to respond to pregnancy," she said.

If the breast needs to be responsive in pregnancy, it's because it's preparing for its big-night out, its very *raison d'être:* breast-feeding. All its 200 million years of evolution and all its individual months and years of construction and signaling and wiring are for this event. Nowhere is the breast more responsive and more conversant and more mind-blowingly intelligent than where there's an actual baby on tap.

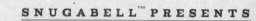

8
WHAT'S FOR DINNER?

First we nursed our babies; then science told us not to. Now it tells us we were right in the first place. Or were we wrong then but would be right now?

— MARY MCCARTHY,
The Group

I didn't bother to read the sections in the pregnancy books about breast-feeding. I was much more concerned about the pain and blood and gore of childbirth. I got stuck on the terrifying bit about pushing a head the size of a bowling ball through what was now bluntly called "the birth canal." I found that part so colossally distracting that I waved off what the books call the fourth stage of childbirth: lactation. I was a mammal. How hard could it be? I would flip through those sweetly illustrated sections later if I made it through the delivery alive.

How wrong I was.

What I didn't know, what I couldn't know, was that childbirth ended up being the easy part. It turns out I was a bit of a champ at it. Nurses filed into my room to watch my breathing technique. In between contractions, they talked about real estate. I didn't need drugs; I didn't even accept an Advil when it was over. "You're tough," said my doctor, shaking his head. My son was beautiful, if a little orange looking. My pride swelled.

But then came the pain and the blood, and it came from breast-feeding, the part of the deal that was supposed to be all saccharine and drenched with love hormones. The first time Ben latched on was wonderful, a little strange, but the fact that he knew what to do seemed a miracle. His strong little mouth created a vacuum like a particle accelerator. The second time he latched on, it hurt, and the third time, it hurt more. My nipples grew inflamed, then formed canyons of fissures, then bled. They looked mangled. I couldn't wear a shirt, much less a bra. My mother-in-law came to visit, and I staggered around the house looking like a crazy bleeding topless person who'd had an unfortunate accident with farm equipment.

I was doing it all wrong. What I learned the hard way is that neither women nor babies "know" how to breast-feed, despite this enterprise being a fundamental part of our humanity. (To be fair, the babies know more than the mothers. Studies have shown that right after birth they are capable of a heroic "crawl" to the nipple, which might be colored extra dark for their blurry-eyed benefit.) If we human mothers once instinctually knew how to nurse babies, we lost it along with things like the ability to make vitamin C. Through our evolving social context, we learned from each other how to eat foods with vitamin C and how to tickle an infant's chin so his mouth will open bigger for breast-feeding. Now, though, we have lost the social transmission that came from living in kin groups. We are replacing it with the paid profession known as "lactation consultant."

Mine was named Faylene, and she made house calls. Friendly but no-nonsense, she showed me the football hold, the lying-down hold, even the upside-down hold (the baby, not me). She helped me open my son's mouth wider and stuff more of my areola in it, and she showed me how to gently break the force-of-nature suction with my pinky when it was time to stop. It was bewildering, but I was getting the knack. Then a relative noticed my son was now even more orange hued. He was diagnosed with a condition called breast-milk jaundice, in which some unknown component of my milk was temporarily interfering with the ability of his liver to break down bilirubin. A pediatrician told us that if this weren't corrected by a twenty-four-hour break from breast milk and immediate application of artificial light to his skin, he would suffer brain damage.

We fed him formula from a bottle for a day and a night while I tried to pump my engorged breasts. When it was time for our reunion, Ben looked at my nipple like it was a foreign metal object. Faylene told me this is called "nipple confusion." I called her back for more body contortions and face stuffing to reacquaint Ben with the real deal. We were finally getting everything sorted out on day 10 when I suddenly felt like I was going to die. My temperature spiked to 104 and my right breast turned to red, hot cement. I went to the emergency room. I had mastitis, a blockage or inflammation of a milk duct that triggers a systemic infection. I needed antibiotics, and I needed them fast. I couldn't help but wonder how humanity had made it this far. What happened to cave women with yellow babies and clogged ducts and no ER? Breast-feeding may have helped the species evolve, but not before killing off a good percentage of its mothers with what used to be called "milk fever."

I would get mastitis three more times that first year. I'm not sure what propelled me to stick it out. Faylene, probably, and a dogged sense of granola-girl duty. But then, once the agony ceased, I found I really liked breast-feeding. In fact, I loved it. Ben and I would settle into our bright-yellow glider at all hours of the day and night. I learned about things that went on along my street at four in the morning that I never imagined. Sometimes I flipped through a magazine or just marveled at my son's now-porcelain skin. I loved the surges of prolactin, a gentle stoner hormone, and of oxytocin, which, as one writer describes it, produces "slight sleepiness, euphoria, a higher pain threshold, and increased love for the infant." I loved the lazy intimacy with my son, and the way he panted and flapped his arms with joy when it was time for dinner.

We were a team. After that first mastitis episode, I gave up pumping. My ducts seemed to clog up too easily, and my son would never take a bottle again anyway. Put in the cold terms of economics, pumping allows women and babies to decouple supply and demand. At the same time, they can decouple themselves, which can be very convenient. If you pump and freeze (as opposed to pump and dump), there's enough for a dry day or an extended absence from your baby, or a chance for your partner to bottle-feed in the middle of the night. Since such arrangements were not to be in our household, we had to be in perfect sync. There was an urgent, immediate need for my breasts to be in good working order. What my breasts cooked up, my baby ate on the spot. When the baby had a growth spurt and fed more frequently, my breasts magically responded by stepping up production. My husband, well rested every night, was just fine with this arrangement.

"Lactation is a father's best friend," he said, heading happily off to work. It's yet another reason for men to like breasts. It's yet another reason for new mothers to want to throttle their husbands.

My pediatrician wore red Converse high-tops and a graying ponytail. When I complained to him in the early months that my son woke up every two or three hours for a feed, he looked at the baby and said, "You little shit." Then he explained that's the way it's supposed to work. Too many parents expect their babies to sleep through the night; it's probably one of the factors driving the use of formula, which takes the baby longer to digest.

I always knew I would breast-feed, and most of my peers would consider it a near felony not to, even though few of us were breast-fed as babies. My mother nursed me for all of four weeks. I know this because I've inherited her journals. There's one labeled, "F Feeds." It was the late 1960s, when only 20 to 25 percent of all American women tried nursing. This era represented the formula companies' strongest headlock on mothers and pediatricians, a time when the science (and profits) of nutritive molecules trumped the fumbling art of breast-feeding. My mother, normally an intuitive spirit with a high disregard for paperwork of any kind, must have been influenced by the men in lab coats. Her journal reads a bit like a high school science log: *Mar 20, 1:15pm: L breast — 15 mins. R breast — 12 mins.* No wonder she gave it up.

By then, breast-feeding rates had been declining steadily since the postwar baby boom, falling by half from 1946 to 1956 as mothers readily turned to the science of the bottle. Today, some breast-feeding advocates — the lactivists or nutricionistas, as they're sometimes called — make you out to be a freak of nature if you don't breast-feed. Historically, though, this is inaccurate. As "natural" as breast-feeding is, there has always been a cadre of women who couldn't or wouldn't do it, for either physiological or cultural reasons. Humans are the only mammal for whom not lactating is occasionally an option (although elephants, foxes, and primates have been known to nurse each other's young). Archaeologists have found four-thousand-year-old graves of infants buried beside ancient feeding vessels lined with residue of cow's milk. (It's no wonder the infants died; for most of our history, not being nursed by someone usually meant a death sentence.) Sometimes mothers couldn't nurse because they had died in childbirth, or their milk dried up after a breast infection, or they were ill. Syphilis, which could be transferred to the baby from the mother during breast-feeding, deterred many in Europe after the Middle Ages. Even fashion posed a problem; the tight corsets of the Restoration were known to flatten or even invert women's nipples. And once the industrial revolution started, many working-class women took jobs away from their homes and babies.

Into these voids marched professional

men with their half-witted and highly politicized ideas. Pliny and Plutarch were opposed to the practice of hiring wet nurses, but Plato loved the idea, "while taking every precaution that no mother shall know her own child." Good thing he stuck to philosophy. Many ancient pundits and doctors offered suggestions for finding the best nurse: she should be cheerful and not deranged, and she should have a strong neck and moderately sized breasts, according to Avicenna in the eleventh century. Babylonia's Code of Hammurabi, circa 1750 BC, was specific about laws and punishments for errant wet nurses: "If a man has given his son to a wet-nurse to suckle, and that son has died in the hands of the nurse, and the nurse, without consent of the child's father or mother, has nursed another child, they shall prosecute her; because she has nursed another child, without consent of the father or mother, her breasts shall be cut off."

Engaging the services of a wet nurse wasn't always the result of dire necessity. The practice was very fashionable in the upper classes of many societies across much of recorded history, probably because it was known to increase the fertility rate of the mother. Prolactin, one of the key hormones triggered by breast-feeding, suppresses ovulation. Nature was wise in this birth-spacing design. Even today, a child born in a developing country less than two years after an older sibling is almost twice as likely to die as a child born after a longer spread. Wet-nursing not only allowed mothers to dodge nature's duties but also resulted in some serious social engineering. While the rich procreated annually, poor women — the wet nurses — often "dry-fed" gruel to their own babies, resulting in huge mortality rates; meanwhile, their own fertility was squelched by their line of work. (This contraception, though, came in handy as some wet nurses moonlighted as prostitutes.)

The natural functioning of breasts has been upended by culture for a long time. While some mothers ducked out of breast-feeding, others were transformed into virtual dairy cows. In Dickensian foundling hospitals established for abandoned babies, wet nurses fed dozens of infants, nursing as often as thirty-four times a day. Sometimes the babies were fed a supplement of rancid cow's milk and flour. The results were grim; death rates in foundling hospitals in the eighteenth and nineteenth centuries ran as high as 90 percent. Even for the urban middle and working classes, who farmed their babies out to wet nurses in the country, mortality rates reached 50 percent. Jane Austen's story was typical. Three months after she was born in 1775, her parents sent her off to the nurse's house, as they had her siblings before her. "When they approached the age of reason and became socially acceptable, they were moved again, back to their original home," wrote Austen's biographer Claire Tomalin. For harried parents, it was a rather appealing if appalling arrangement: pop the children out, send them off packing, and bring them back when they're big enough to do some chores. Some sources claim this is where the term *farmed out* derived.

Dry-feeding, despite its disastrous results, was also rather common because it was cheaper than hiring a wet nurse. Many mothers throughout the ages suffered either real or perceived bouts of insufficient milk, and so they would resort to feeding babies some sort of mash as a milk supplement or total replacement. (Breast-feeding is an ingeniously calibrated feedback loop, and it's not always forgiving. Once you embark on the slippery slope of supplemental feeding, milk production can taper precipitously.) Many recipes were handed down the generations for the ideal food, usually involving some sort of milk, water, grain, and sugar. Occasionally wine or spirits, cod liver oil, and opium were added. Before the days of refrigeration and pasteurization, such cocktails were at best a dicey proposition for infants, who have immature immune systems. Of the dry-fed kids who survived, many had scurvy, rickets, and iron or other mineral deficiencies.

It's not surprising that by the end of the nineteenth century, medical professionals, who had increasingly elbowed their way into the realm of midwives, would also turn their attention to infant food. Motivated by high mortality rates due to malnutrition and intestinal disease (and seeking to cement their own job security), leaders in the field actively recommended that doctors replace "old women" and "uneducated nurses" when it came to overseeing the infant diet. In fact, the growth of the pediatrics field at the turn of the twentieth century was predicated on infant nutrition. In 1893, the medical lecturer John Keating called it "the bread and butter" of the profession. These doctors tinkered with their own formulations for milk substitutes and readily experimented with offerings in the growing marketplace. Mothers had to visit the doctors regularly to get access to the food, which was available only by prescription. Together, the formula companies and the doctors reinforced each other's businesses.

Two major developments around this time nudged the creep away from breast-feeding. One was the rise of manufacturing, which allowed formula companies to produce large amounts of relatively consistent product. Henri Nestlé was a young chemist and merchant surrounded by dairy cows in Vevey, Switzerland. In 1867, he concocted his Farine Lactée Nestlé. He described his Milk Food as "good Swiss milk and bread, cooked after a new method of my invention, mixed in proportion, scientifically correct, so as to form a food which leaves nothing to be desired." By the 1870s, the product had gone global. (Now, 140-plus years later, formula has propelled Nestlé to being the largest food company in the world.)

The other major event was the rise of germ theory. It changed modern medicine forever, and much along with it. In a nutshell, this was the recognition that microbes caused disease. Before that, people thought sickness was caused by swamp vapors or spontaneous eruptions or a combination of bad luck, bad behavior, and God. In 1876, the German physician Robert Koch proved the bacterium *Bacillus anthracis* caused anthrax in livestock, and he discovered the bacterium that causes tuberculosis. Over the next twenty years, microbiologists isolated the organisms that cause pneumonia, diphtheria, typhoid, cholera, strep and staph infections, tetanus, meningitis, and gonorrhea. In one sense, an environmental understanding of disease was being replaced by a germ-based one. It represented medical progress, but it would ultimately provide a false sense of separation from nature.

The new discoveries led to many vaccines and antibiotics, as well as to vastly improved sanitation measures, quarantines, and food safety practices, such as pasteurization. Many, many lives were saved. You might think an understanding of deadly bacteria would bolster the argument for breast-feeding and against commercially processed infant foods, but the opposite happened. Medicine and science enjoyed a new prestige, and mothers grew more willing to surrender traditional knowledge and personal control over infant care. The midwives and grandmothers didn't stand a chance. In 1920, only 20 percent of American women gave birth in hospitals; by 1950, over 80 percent did. There, scientific motherhood flourished.

Such science did not look favorably upon

breast-feeding. Mid-century mothers were often totally anesthetized for childbirth, necessitating the use of forceps to pull the baby out. After birth, babies were typically given formula to wait out the period of time before the mother's milk came in (around three days, but before that she makes immune-boosting colostrum). Babies and mothers were separated at birth, only to be reunited for short, closely regulated, and highly sanitized feedings. Talk about some weird amendments to our eons-old mammalian patterns: mothers wore masks, scrubbed their nipples with soap, nursed their babies, and then watched them get burped through Plexiglas windows. To allow the mothers to rest, nurses took over the night feeding with formula freely supplied by manufacturers. Only several other feeds were allowed during the day, usually with one breast at each feed. It's no wonder the mothers didn't produce enough milk. The babies became hungry, and the bottle was presented as a perfectly acceptable alternative. The doctors and nurses had little-to-no training in lactation but much expertise in formula measurements. After a week of the Nurse Ratched approach to baby care, mother was sent home with a pat on the back and a free sample of Nestlé's finest.

In 1956, a small backlash ensued. It started in Illinois at a church picnic with a couple of Catholic matrons, not normally known for upholding mammalian urges. "It just didn't seem fair that mothers who bottle feed . . . were given all sorts of help . . . but . . . when a mother was breast-feeding, the only advice she was given was to give the baby cow's milk," said Marian Thompson. She and Mary White formed a small group to support other breast-feeding mothers. They called themselves the La Leche League. As one of the founding members told the *New York Times,* "You didn't mention 'breast' in print unless you were talking about Jean Harlow."

You probably know the tale from here. The suburban La Leche ladies went on to align with the hippies and the back-to-landers, and together they reformed hospital practices and reversed the sorry downward spiral of breast-feeding rates. Then they took on Nestlé through an epic boycott over its capitalist hegemony in developing countries, where infants were dying from formula made with contaminated water. The league is alternately inspirational and infuriatingly dogmatic. For a time, it advocated that mothers stay home and not work. This proscription did not sit well with the emerging feminist movement in the 1970s, and tensions over their take-no-prisoners attitude remain to this day. Still, the organization has effectively hammered home its central message that breast-feeding makes babies smarter, healthier (those ear infections!), less obese, more loved, and pretty much superior in every way.

The lactivists, both nationally and globally, have made their milky mark: the World Health Organization now recommends breast-feeding for two years; the American Academy of Pediatrics recommends it for one. Many hospitals still distribute free formula (mine did), but they also allow "rooming in" so new mothers and babies can spend all their time together; often, baby is put to breast within moments of delivery.

Even so, American mothers learning to breast-feed are uniquely beset by problems. It normally takes about three days after childbirth for a mother's milk to "come in." Any longer than that, and the docs are going to whisk Junior away and start formula. So the early days are critical for the long-term success of nursing. In Ghana, only 4 percent of women experience delayed lactation; in Sacramento, site of a recent study, 44 percent do. No one knows why. It might be because we're more obese or older, or we use spinal anesthesia, or we have more C-sections or more environmental exposures, or tight bras have flattened our nipples.

Stay tuned; it's an active area of research.

Despite the lactivists' best efforts and deepest intonations of "breast is best," breast-feeding rates here and in much of the world remain middling. In the United States, about 70 percent of mothers initiate breast-feeding, but only 33 percent nurse longer than six months and only 13 percent fulfill the year-long recommendation. Australia and Sweden, with their generous maternal leave policies, have gone lacto-manic, with initiation rates of over 90 percent. Canada's is about 87 percent. Brazil has been a great public health success story: The average duration of breast-feeding has increased from two and a half months in the 1970s to over eleven months today. Over 95 percent of Brazilian women try breast-feeding, and for those who can't swing it, the country has two hundred human milk banks and over 100,000 donors. Their milk is collected and stored by firefighters.

Has all the ruckus — all the maternal guilt, the physical and mental introspection, the *madre-a-madre* name-calling, the battles with the medical establishment — been worth it? Is milk, *au naturel,* really so superior to formula that we must make each other feel bad about our failures and choices? The honest answer to this question is yes and no. I don't mean to be feckless. Breast milk confers many known benefits to infants, but for healthy babies in the developed world, those benefits are relatively small. They might actually be bigger than we think, but the truth is, we don't really know. To some extent, the activists asserting we're in the midst of "a biocultural crisis" have perhaps overstated their case.

The journalist Hanna Rosin wrote a compelling essay challenging breast-feeding in a much talked-about *Atlantic* article in 2009: "And in any case, if a breast-feeding mother is miserable, or stressed out, or alienated by nursing, as many women are, if her marriage is under stress and breast-feeding is making things worse, surely that can have a greater effect on a kid's future success than a few IQ points. . . . So overall, yes, breast is probably best. But not so much better that formula deserves the label of 'public health menace,' alongside smoking."

It was a strong argument delivered well

for an *Altantic* readership. Who needs a few extra IQ points or a few less visits to the doctor when our children will be high achieving and successful anyway? Given the context, it's a reasonable stance. And refreshing, a welcome astringent to the sanctimony of the lechistas. I'm sure many mothers read it and gleefully ran out to the market — perhaps Whole Foods in their case — to buy Earth's Best Soy Infant Formula. After all, most of us weren't breast-fed and look at us: we're healthy, long-lived, long limbed, and terribly clever with iPhone apps.

Then again, once you look outside Rosin's reader demographic, a lot of us are rather obese. And diabetic. On the road to heart disease? Check. And because of all this, some of us will be facing shorter lifespans than our parents. Allergies and asthma? Common. (And a word about the IQ scores: formula confers the same average loss in points — four — as high childhood lead levels. That dip was enough to trigger a public health outcry in the 1970s, and resulted in federal laws banning lead in gasoline and paint. Now kids' scores are back up.) But can we attribute any of our metabolic malaise and chronic ill health to formula? Some people are trying, and they're trying hard. So far, the data are intriguing, but underwhelming.

Many studies, for example, have compared formula and breast-feeding for a risk of obesity in infants and children. The results of these studies, it must be said, are all over the map. Two major reviews of the literature found that in a majority of studies, children who were breast-fed faced a lower risk — ranging from about 10 to 25 percent — of obesity. Despite the fact that most of these studies rigorously controlled for factors such as the mother's level of education, smoking, and so on, other confounding factors might exist, so it's difficult to know for sure how much of the benefit to attribute to breast-feeding. It is known, however, that formula-fed babies consume about 70 percent more protein than their peers, and this may trigger higher levels of growth factors and insulin secretion, which in turn lead to increased deposition of fat.

Since it's hard to see big bangs from breast-feeding in countries where kids are generally healthy and well taken care of, I wanted to look elsewhere. The data start to get really electric when you peer farther over the fence into areas of poverty or the world of preemies and sick babies.

I owed it to breasts to check it out. So I headed for Peru.

9

HOLY CRAP:
HERMAN, HAMLET, AND THE
ALL-IMPORTANT HUMAN GUT

O, thou with the beautiful face, may the child Reared on your milk, attain a long life, like The gods made immortal with drinks of nectar.

— SUSRUTA SAMHITA,
fourth to second centuries BC

Lima hosted the fifteenth meeting of the International Society for Research in Human Milk and Lactation during a cool austral spring week in October. The society's cochair, Professor Peter Hartmann, welcomed me heartily. "We've never had a journalist before! Maybe you can tell people about our work!"

Hartmann, an Australian now in his seventies, is perhaps the world's foremost expert in lactation. Even so, he bears the demeanor of someone whose work is largely unacknowledged outside this crowd. He's a little bent over, and quiet, and a bit harried. He spent the week in Lima clutching his brief-case and scurrying from meals to meetings wearing his beret and a leather jacket. His son, Ben, is also here, not quite as stooped or as sartorial. Ben, thirty-four, runs a milk bank, collecting and storing donated human milk for use in the preemie ward of King Edward Memorial Hospital on the edge of Perth. (What is it with fathers and sons dove-tailing careers around breasts? Ben's infant, Arlo Peter, has actively ben-efited from the Hartmann male tradition. "As my poor wife has indeed been subjected to the full barrage of tests by the research group, we certainly had a lot more informa-tion at our fingertips than most," said Ben.)

I sat down with the elder Hartmann dur-ing one of the numerous café breaks in a heavily tiled, open-air meeting room at our faux-Renaissance lodge. Originally, he told me, he intended to study dairy science. He got a Ph.D. in bovine lactation. But then Britain changed its export policy, and the Australian dairy market "disappeared over-night," he said. He secured a lectureship in biochemistry at the University of Western Australia in Perth, and in 1971, his first child was born. That event piqued his inter-est in the human side of things. He started studying progesterone withdrawal in women after birth, and found a large pool of enthusiastic breast-feeding volunteers through the Australian version of La Leche League. Human lactation was a tough academic sell, though. "Nobody was really interested when I applied for grants. It wasn't a good career choice." He smiled impishly, then added, "I proved them wrong."

Still, he said, "It's amazing how few people are interested in this incredible organ. The breast is the only organ without a medical specialty. It represents 30 percent of a woman's energy output, and it's not represented by a specialty! It's absolutely appalling!" What he meant by the energy bit is that while a woman is lactating, the metabolic energy required to feed her infant is 30 percent of her total output — or the energy equivalent of walking seven miles — every day. Looked at another way, a male baby requires almost 1,000 megajoules of energy the first year of life. That is the equivalent of one thousand light trucks moving one hundred miles per hour. As the ecologist and writer Sandra Steingraber has put it, "Breastfeeding is a form of matrotropy: eating one's mother." No wonder so many women are ambivalent about doing it.

"It's a magnificent organ to study from a

molecular standpoint," Hartmann continued, occasionally smoothing his trim white beard. "It's easy to get access to it and harvest molecules. The problem lies with the view of it as an aesthetic breast. You only have to go to the local newspaper and see breasts hanging all over the place." (Hartmann must be one of the few Western men on the planet who see this as unfortunate.) "The problem is the view of the aesthetic breast gets in the way of the view of the breast-feeding breast. The guys at the tennis club joke they wish they had my job, but no one is *doing* my job. At other [biological] meetings, you see thousands of scientists. We've got less than a hundred."

He's right that breasts are often overlooked, at least for non-cancer scientific research. The Human Microbiome Project, for example, is decoding the microbial genes of every major human gland, liquid, and orifice, from the mouth to the skin to the ears to the genitals. It neglected to include breast milk, the life-giving, life-saving, older-than-mammals-themselves elixir. Oops.

There is at least one entity very interested in Hartmann's research, and that is Medela, the Swiss maker of breast pumps. The company sent a number of representatives to Lima, and they presented a poster explaining their latest product. It's a new artificial "teat" — that's Australian for *nipple* — based entirely on Hartmann's lab studies regarding flow and suction. I know a thing or two about suction. Just talking about it makes me wince. The new teat is meant to be used with Medela bottles filled with pumped human milk.

The Hartmann lab in Perth is well regarded for upending prior notions of how sucking — technically, *suckling* — works. Experts used to think the infant squeezed the nipple with her tongue, rhythmically releasing milk through this peristaltic action, a bit like wringing a washcloth. But Hartmann and his colleagues, using high-tech ultrasound videos, showed that the baby forms a strong suction with her lips, and it's when she *releases* the nipple that milk flows down her throat (and moreover, this being a specialized human infant, she can suck and breathe at the same time, unlike adults).

One day some years ago, Hartmann was flying over Australia, and he found himself gazing out at giant ore piles near the mines. They were large mounds of salt and minerals, smooth and gently rounded. They looked like . . . his favorite organ. From high in the sky, a stockpile looked just like a breast from a few feet away. It occurred to him that it might be possible to apply giant-earth stereoscopic measuring techniques to the human breast. But he wasn't merely interested in measuring the volume of a breast; he wanted to measure the synthesis of milk.

"Humans do not produce milk at full capacity, like a dairy cow," he said. "They down-regulate to match the baby's appetite. So we had to learn about those differences. How do you measure milk synthesis in a woman? I thought maybe if you could measure the volume of the breast, you could measure synthesis." So he approached an expert in a mine-measuring technology called Moiré topography, and together they figured out how to calibrate units in something other than metric tons. They call it CBM, for computerized breast measurement, and it has to do with projecting light stripes at an angle onto the breast. "The distortion of the stripes could let you work out the volume of breast!" said Hartmann. "We could do it before and after feeds over a twenty-four-hour period. The difference in volume is the short-term synthesis [of milk] from one breast-feed to another!"

In the old days, people used to measure

milk output by simply weighing the baby before and after a feed, but that didn't reveal information about the workings of each breast independently, or about how much milk a breast could make from hour to hour or day to day. These data could be useful to hospitals, doctors, and, of course, Medela, which funded the research. In a paper describing the work, Hartmann and colleagues found that each breast of the average new mother produces approximately 454 grams, or 16 ounces, every twenty-four hours. Each breast can store about half of that, and both actions are determined by the demand of the infant (one baby in the study ate twice the average). Check this out: even after fifteen months of lactation, each breast can still make 208 grams of milk, even though the breast has *returned to its pre-pregnancy size.* In other words, the breast becomes more efficient, possibly owing to a "redistribution of tissues within the breast," according to the paper by Hartmann's lab. Breasts should get an Energy Star rating.

In any case, thanks to Hartmann, no longer is the dairy machinery such a mystery.

But the dairy *product* still is. I wanted to find out more. What makes milk so special, if it really is? A lot of Hartmann's work is in the liquid physics arena, but many of the Peru attendees were molecular biologists, biochemists, or geneticists who are deconstructing the components of milk bit by bit. They've been doing this for well over thirty years, and you'd think they'd have figured it out by now. Until very recently, it was thought that breast milk had around two hundred components in it. These could be divided into the major ingredients of fats, sugars, proteins, and enzymes. But new technologies have allowed researchers to look deeper into each of these categories and discover new ones altogether.

Scientists also used to think breast milk was sterile, like urine. But it turns out it's more like cultured yogurt, with lots of live bacteria doing who knows what. These organisms evolved to be there for a reason, and somehow they're helping us out. One leading theory is they act as a sort of vaccine, inoculating the infant gut so it can recognize bad actors and fight them when the need arises. At the conference, Mark McGuire, another former dairy scientist recruited to the human lactation field, described how he took forty-seven samples of milk, extracted DNA, and identified eight hundred (yes, eight hundred) species of bacteria living there, including small amounts of staph, strep, and pneumonia, all of which normally live in our bodies. An individual milk sample has anywhere from one hundred to six hundred species of bacteria. Most are new to science.

Then take the sugars. There's a class of them called oligosaccharides, which are long chains of complex sugars. Scientists have identified 140 of them so far, and estimate there are about 200 of these alone. The human body is full of oligosaccharides, which ride on our cells attached to proteins and lipids. But the mother's mammary gland cooks up a unique batch of "free," or unattached ones and deposits them in milk. These are found nowhere else in nature, and not every mother produces the same ones, since they vary by blood type. Even though they're sugars, the oligosaccharides are, weirdly, not digestible by infants. Yet they are a main ingredient, present in milk in the same percentage as the proteins and in higher amounts than the fats. So what are they doing there?

They don't feed us, but they do feed many different types of beneficial bacteria that make a home in our guts and help us fight infections. In addition to recruiting the good bugs, these sugars prevent the bad bugs from hanging around. They act as "anti-adhesives," kicking the bad guys off the gut surface. Some also seem to handcuff themselves to the criminals and escort them off the premises like a micro paddy wagon. "I think the benefits of human milk are still underestimated," said Lars Bode, an immunobiologist at the University of California, San Diego. "We're *still* discovering functional components of breast milk using new technologies and using smaller amounts of milk."

Bode told me it's well established that premature infants do remarkably better — as in, an order of magnitude better — on breast milk than on formula. As we've been able to keep younger and younger preemies alive, they're more likely to be very sick. About 10 percent of preemies will suffer from a dreadful disease called NEC, or necrotizing enterocolitis, and about a fourth of those will die from it. NEC is a gut infection that causes the lower intestine to shrivel up and die. Babies who survive this often must have the necrotic portion surgically removed, leaving them with a condition called short-gut syndrome. Because they can no longer adequately digest food, they spend the rest of their lives attached to an IV. *But the incidence of NEC is 77 percent lower in breast-fed babies than in formula-fed babies.* This is why neonatal units work so hard to get mothers to pump breast milk for their preemies' feeding tubes. In lieu of that, they use donated milk from milk banks, and in lieu of that, they can buy a newfangled "fortifier" made from concentrated *human milk* by a company called Prolacta Bioscience. It costs $12,000 per baby.

Naturally, conventional formula companies are falling all over themselves to synthesize these unique human sugars and add them to their cow-milk products. So far, they've been able to re-create only a few of the simpler ones, and to not much effect. These newly enriched formulas, for example, do not alter preemie NEC rates, according to Bode. This is because the most "bioactive" molecules are the bigger, more complex oligosaccharides, which are incredibly difficult to make in a lab. "Take one of these special monosaccharides? If you wanted to supplement human milk with it, the package would cost half a million dollars," he said.

Bode said his lab has also shown that a simpler chain, called GOS, is effective at fighting amoebiasis, a parasite that kills 100,000 people a year. It's likely GOS could help adults as well as babies.

The gut, it turns out, is incredibly important not just to infant health but to adult health as well. Bruce German, a food chemist from the University of California, Davis, drove this point home for me. Picking up the breast ball where the Human Microbiome Project dropped it, German is spearheading the Infant Microbiome Project at the university's Foods For Health Institute. The idea is to map and characterize oligosaccharides, as well as other human milk components, for eventual infant and adult medical applications. As German put it in a recent video, "We'll take little tiny droplets of milk and disassemble them completely, understand every single molecule in it . . . and how they function when ingested by infants. . . . We're confident it will teach us how to prevent diseases like diabetes and heart disease and ultimately even cure diseases like cancer."

German speaks in superlatives and alluring metaphors even when he's not being filmed. "The genomic tree of life is a compelling one," he told me in a private mini-lecture on microflora over breakfast one morning. A wiry, almost hyper speaker, he began to turn red. "We're just a tiny branch! The rest of it is all microbial! We're just a small part of the mass of microbial life. To a large extent, we live our life at their say-so. We have to form a pact with the world around us. Recruiting protective microflora is the very first thing we do in life." He wasn't eating, but now he sipped his coffee. "Put yourself in the mindset of an infant. You're born, you are dropped in the mud, literally, where the microbial community thinks of us as lunch. So you must develop a community that protects you for life."

He continued in his riveting, pretend-you're-a-baby mode. Clearly he's a man used to speaking to glassy-eyed undergraduates. "If you're a preemie and you probably came out by C-section and you're not breast-fed, you're acquiring bacteria in your gut from the hospital that will reside there for the rest of your life." He winces. "That's not the way you want to do it. Normally, the mother hands down bacteria to her infant by means we still don't fully understand. If there's a successful transfer, it will be handed down mother-to-daughter for generations. Now we think one C-section could break that chain." He plunks down his coffee cup. "You've lost touch with your genetic ancestry."

Could another way to break the microbial chain occur if the mother herself was not breast-fed? Are we now looking at a couple of generations of orphaned intestines, cut off from their full bacterial legacy? I asked German and he nodded. "Exactly." He said he'd like to see every baby (who doesn't receive breast milk) get a dose of *Bifidus infantis* at birth, like a vitamin K shot.

This bacterium is one of German's favorites. *Bifidus infantis* has seven hundred genes, all of which evolved to live and thrive in a unique microbial environment: the infant gut. There, *B. infantis* eats the oligosaccharides that rain down in breast milk. The bacterium produces proteins that pull those special sugars inside it, where they get broken down and digested so that they are unavailable for other (worse) bacteria. Furthermore, said German, "*Bifidus* can overwhelm bad bacteria. It recruits a whole schoolyard of supportive organisms." As I've said, these human milk oligosaccharides are not found anywhere else in nature. "It's clear that these bacteria co-evolved with our oligosaccharides," continued German. "It's our true symbion."

It's good to know all this, but it's also another intense failing for modern mothers to feel guilty about. I couldn't help but wonder if I'd somehow flubbed the important microbial hand-off to my children. Some of the blame I could cast to my mother, who might not have given me the adequate goods in just four weeks of nursing. But I get a pit in my stomach when I think of the repeated rounds of antibiotics I took for mastitis while nursing my son: cephalexin, amoxicillin, dicloxicillin. Was I killing off everything good in his gut as well as in mine?

The adult gut can recover from antibiotics within a few months because of mysterious reserves of bacteria in our bodies, our housemates, and to some extent, our food. But the infant, who is bacterially "naive" and building colonies for the first time, might not. These thoughts snowballed in my mind, because my son has had gastrointestinal trouble for most of his life. The poor guy is chronically constipated. One pediatric gastrointestinal specialist I took him to shrugged off a search for root causes, saying, "Well, some people just have slow motility, like a sloth." That's my son, the sloth.

I sought out David Newburg, a biologist from Boston who has been studying the connections between breast milk, intestinal microflora, and disease for over two decades. A tall, tanned man with a trim goatee, I often saw him enjoying the Peruvian pastry table, as did I.

"Can I ask you a personal question?"

Newburg raised his eyebrows. "In that case, I'm going to need two tea sandwiches," he replied.

I told him the story of my mastitis and the antibiotics. Was my son's problem my fault?

"It's definitely possible," he said, making me feel wretched. Good tests to diagnose the gut's array of normal and abnormal microflora are still a few years out, he explained. Changing the microflora is even harder. Someday, though, these things will be a routine part of medical care. In the meantime, Newburg recommended Ben regularly take probiotics (such as the lactobacillus found in yogurt and supplements) as well as eat foods rich in prebiotics (the complex carbohydrates that beneficial bacteria need to thrive). Although breast milk is the world's best source of prebiotics for humans, they can also be found in Jerusalem artichoke, Belgian endive, onions, asparagus, and some other plants not terribly alluring to a nine-year-old.

I told Newburg what Bode had said about it being so difficult and expensive to synthesize human-milk prebiotics.

"Hah!" said Newburg, baring a little professional rivalry. "We know how to do it. Come visit my lab. We're up to our necks in shit."

I know, most people would have declined this offer and fled. But by now I had fallen too far down the milky rabbit hole and was weirdly entranced by gut flora, that unexpected and invisible pillar of human health. I kept thinking of German's image of us living at the behest of the microorganisms, not the other way around — "Who's really cultivating whom?" he'd asked dramatically. The microflora outnumber us by a lot. There are ten times more microbacteria in our guts than there are cells in the human body. A song lyric kept playing in my head: Patty Griffin's "You are not alone."

And so, a few weeks later I was negotiating the steep, rain-slicked steps outside of Higgins Hall on the Boston College campus. I maneuvered past a rather severe statue of St. Ignatius and into the new and immaculate molecular sciences building, where Newburg commands a spacious realm on the fourth floor.

Wearing black jeans, a black golf shirt, and sandals, Newburg welcomed me into his lab. It looked like a cross between a kitchen and a Kinko's. The boxy, beige machines are actually mass spectrometer contraptions, such as the snazzy new "triple quad." It sounds like a Vail chairlift and looks like a photocopier, but it costs around half a million dollars and breaks down molecules into gradually smaller components. To distinguish and identify different molecules, these machines utilize color, molecular weight, or, my favorite, "time of flight." This one sends molecules pinging down a zigzag chamber and then up a small cylinder the size of a stovepipe. No two long-chain molecules make the lap (or, technically, have the same mass-to-charge ratio) in exactly the same way. Many of the substances Newburg is finding in human milk have never been seen before.

The lab is a bank for two main kinds of substances: human diseases and the breast milk that fights them. To obtain the disease-causing organisms, Newburg collects infant feces. He and his colleagues isolate the pathogens (such as botulinum, campylobacter, *Vibrio cholerae,* and *Escherichia coli*) and grow them in an anaerobic chamber similar to our guts. He especially treasures a source in Mexico, a clinic that sends him samples rich in things like rotavirus. Otherwise, he finds them through local hospitals and lactating-mom networks. "Just to handle baby poop is an incredibly long and complex process," he said, involving informed-consent paperwork and hospital review boards. Some of his fecal freezers are set to −80 degrees Celsius, the temperature of outer space. Other incubators mimic body temperature for growing human cells from the lining of intestines and lungs (breast milk is also ferociously adept at fighting pneumonia). Leaving the tissue culture room, I saw a tube the size of a beer glass stuffed with what looks like raw steak. "That's a liver," said Newburg.

For him, analyzing baby shit is practical and urgent. Globally, 1.4 million children under five die each year from diarrheal illnesses. This makes sense if you consider that 20 percent of the world's population doesn't use any sort of toilet, and nearly half doesn't have access to decent sanitation. Nearly a billion people don't live near clean drinking water. At the same time, human milk is so effective at fighting infections that if all children were exclusively breast-fed the first six months of life, one in five childhood deaths could be prevented.

"Breast-fed poop doesn't smell too obnoxious," Newburg said. "It's more like sour cheese or milk. Frankly, even as a guy, I got used to it." Newburg led me to a normal-looking fridge to show off some of his precious collection, but he gasped when he saw the door was slightly ajar. A box of test tubes was wedged clumsily into the door. A small puddle had formed on the floor below it. "Oh no," he said. A lab tech had accidently left the door open overnight. It occurred to me that the only thing worse than a freezer full of poop was a freezer full of thawing poop — especially for Newburg, who would have to deal with the scientific consequences. He lifted a test tube packed with brown goo and shook it. "I think this whole fridge is compromised," he muttered.

After he washed his hands, we settled into his adjacent office for a chat. A poster-sized, soft-focus photograph of a blonde woman nursing a baby loomed above his desk ("My wife doesn't like it," he said of the image). Books such as *Phospholipids Handbook, Gray's Anatomy,* and *Modern Nutrition in Health and Disease* packed the front wall. Like many men in this field, Newburg told me he didn't start out intending to study lactation. His field was neuroscience. But running a rat experiment three decades ago, he noticed his formula-fed pups "never performed as well as the nursed ones. A normal person would have said, 'fine,' but not me. I took a sabbatical to study essential nutrients for brain development," and the rest is history.

He became intrigued by the indigestible oligosaccharides, and soon he had established that they must function to fight pathogens in the infant gut. In the 1980s, his lab (then at Harvard) rather startlingly discovered that human milk inhibits the transmission of HIV, among other things. He didn't know exactly how, and he still doesn't, although he's closer to knowing which oligosaccharide compound is responsible. "We do know that the transmission of HIV through milk is much less than through any other medium," he said. He fully expects to identify the heroic sugar complex, then make it and offer it up as a therapy in the real world. "We'll study it and we'll find out," he said. "It would be much more effective than a vaccine, I think."

Already, Newburg's company, Glycosyn (he cofounded it in 2002), is making a "2-linked fucosyloligosaccharide" known to help ward off norovirus, *E. coli,* cholera, and campylobacter. Because, as Bode pointed out, it's too expensive to synthesize these molecules from scratch, Newburg has a different strategy. He's teaching yeast to produce them for him by converting a natural product they make into a building block called fucose. He then takes that and links it to lactose "because that's what mom does." Some other companies in Europe are making oligosaccharides from plants or cow's milk and putting them in infant food, but Newburg says it's not the same.

Glycosyn will start testing its product in humans sometime in 2012 or 2013. Newburg told me the final product will probably resemble a sugar packet that can be mixed into food or formula. It will be ideal for babies on formula or babies and toddlers who are weaning, which can be a treacherous process in developing countries with unsafe food and water. Newburg's product will be like NutraSweet for the survival set, the mysterious stuff of breasts purified into a paper packet.

As someone who extols the benefits of breast milk but wants to improve formula, Newburg has garnered some criticism from both sides. If there's one thing the lactivists hate, it's better formula, because they think it can never really be good enough. "It's frustrating to see moms who don't breast-feed, but I understand why some don't," said Newburg. "I don't think their children should be punished. My orientation is to the baby."

■ ■ ■ ■

Lechistas, prepare yourselves: formula will get better and so will a bunch of other foods, supplements, therapies, and medications thanks to the unlocked secrets of milk. A quick survey of what other biotech companies are doing shows the range of benefits being urgently, greedily, attributed to human milk. It's important to remember from chapter 2 that lactation likely evolved from the immune system; its primary function was not nutrition but protection. Most of the cells in milk are macrophages, which disable viruses, fungi, and bacteria. I already mentioned Prolacta Bioscience, which is concentrating and pasteurizing donated human milk — and then selling it — as an "immunonutrition" supplement for preemies weighing less than 2.5 pounds. In the Brave New World department, several companies are reengineering other animals to produce the unique ingredients of human milk, because it's still easier and cheaper to raise a herd of transgenic goats than it is to beg or buy large quantities of milk from suburban mothers.

One of the most sought-after components of human milk is a glycoprotein called lactoferrin. Known to have keen anti-inflammation, antioxidant, and anti-infective properties, it's an iron-binding machine that outcompetes pathogens. Lactoferrin can also be found in tears and saliva and genital secretions, but in tiny percentages compared to milk. It's possible to inject animal embryos with the human gene that makes it. Some companies are genetically altering cows, goats, and even rabbits, then isolating the human lactoferrin from the milk. One Japanese company has begun marketing capsules, which it calls "Lactoferrin Gold." Three *liters* of modified cow's milk are needed to make one capsule. I can see why they named it after a precious ore. To make lactoferrin, another company bred a whole herd of cows from one long-dead transgenic bull named Herman. But altered mold fungus can make it too. A biotech outfit with over a hundred lactoferrin patents intends to use the fungal product for fighting cancer and healing wounds.

According to one economic analysis, if lactoferrin were added to infant formula, it would create an extra $15 billion in value. If added to eye drops, oral hygiene, soaps, and shampoos, another $10 billion. Cancer drugs: $19 billion.

That's just lactoferrin, but there is also

active research on other components. Stem cells, for example, teem from human milk, particularly from the dense colostrum produced in the early days of nursing. Before the baby is five days old, she'll receive five million stem cells from the mother. No one really knows why. Are they colonizing the baby in case she needs them? Are they just a by-product from the newly functional mammary gland? Then there's a very cool protein called alpha-lactalbumin. In the acids of the infant stomach, the protein refolds itself and picks up a neighboring fatty acid, also from the milk, forming a new complex. The scientist who discovered it fifteen years ago, a Swede named Catharina Svanborg, dubbed it HAMLET, for *h*uman *a*lph*a*-*l*actalbumin *m*ade *le*thal to *t*umor cells.

This HAMLET ditches the pretty soliloquys and dons a superhero cape, diving into the nuclei of malignant (and viral) cells and freezing the gears. It effectively prevents malignant DNA from replicating and then in a final grand stage gesture causes the cells to implode. Weirdly and auspiciously, it seems to destroy only bad cells, leaving the good ones alone. Laboratory experiments have shown that HAMLET kills forty different types of cancer cells in a dish, including those of the bladder, lymphoma, skin, and brain, but it has not been tested much in humans yet. Still, the reason Svanborg began looking at milk is that several studies found that formula-fed children have significantly higher rates of childhood lymphoma than their breast-fed peers.

All this ought to make nursing mothers feel a little more valuable. Maybe they'll wise up and start registering with insurance companies as health-care providers. Or maybe they'll join the raw-milk underground. A few already have: an Internet site called Only The Breast lists classifieds with wording like "scrumptious mommy milk." At four dollars per ounce, it costs 262 times the price of a barrel of oil. The marketplace for human milk in most of the United States and in the rest of the world is unregulated so far, despite the fact that it's capable of transmitting hepatitis and other maternal diseases along with lactoferrin. Nonprofit human milk banks (there are eleven in North America) heat the milk to pasteurize it, but the process also kills some of its bioactive ingredients. While donor milk is used mostly in neonatal intensive care units for preemies, older children and adults sometimes buy it for treating various illnesses or for soothing the harsh mucosal effects of chemotherapy.

As Bruce German had reminded me in Peru, these breakthroughs in the understanding of milk are not just interesting; they are fundamentally altering what we know about human health. "The story that is compelling to me is the fascinating interplay between bacteria and humans," he'd said. "This whole story is part of a revolution in science itself, where the chemistry-dominated science of the twentieth century is giving way to the biology-dominated science of the twenty-first century. Such a shift is sometimes difficult to appreciate, especially for people outside the scientific community. For milk and bacteria there is an easy point of entry for people to see the vivid contrast: twentieth-century chemistry — use chemicals to kill all bacteria; twenty-first century — use biomolecules and organisms to guide a supportive microbial ecology. It's a new world of science."

Nature has designed milk to be ingenious, but it's the breast itself that directs the show. Biologists bat around the concept of "crosstalk," how one part of the body communicates with another and vice versa. In the case of the lactating breast, the organ is communicating not only with its immediate landlady but also with the infant. From the very beginning, the breast appears to know whether the infant is a boy or a girl, at least in rhesus macaque monkeys, which have similar milk to humans and have been studied more comprehensively than their human relatives. In the macaques, mothers of sons produced fatter, more-energy-dense milk. Katherine Hinde, a professor in the human evolutionary biology department at Harvard University, thinks this might be because macaque males have slightly higher growth rates and as adults weigh about 30 percent more than females (human males weigh about 15 percent more than females). But Hinde has another, more devious social theory as well. She discovered that macaque mothers produce fattier milk for sons, but they make *more* milk for daughters, meaning the maternal energy investment is about the same. In that matrilineal primate society, daughters learn from hanging around their mothers longer and more often, and thinner milk means they stay close for more frequent feedings. The sons, by contrast, might be "tricked" by the mother's fattier milk into feeling sated and therefore not feeding as often. It's not a bad thing for the sons; they have more time to play and explore, skills they'll need down the road when they leave the group.

How does the breast know whether the infant is male or female? Probably because hormones called placental lactogens talk to the breast during pregnancy, when it is building the structures it will need for making milk. Girls evidently get the skim-milk machinery. Mom wants to produce good milk, but she doesn't want to kill herself doing it, which brings up an interesting dynamic between mother and infant: competition. Babies have evolved their own tricks to get as much of their mother's resources as they can: Witness the tightly interwoven placenta (made by the embryo) that becomes essentially parasitic. The mother's body has genes to expel the fetus a little before the due date. The baby's genes — put there by the father, presumably — tell it to stay put a little longer.

The breast picks up the tug-of-war after birth, slipping endocannabinoids into the milk. Note the root *cannabis* in there. These substances, which cause the munchies, probably play a role in enticing infants to eat. But they also regulate appetite so infants feel *very full* by the end of a feed and thus don't eat too much. Interestingly, formula lacks these compounds, and formula-fed babies have a notoriously higher caloric intake. It's one of the speculations about why we have a childhood obesity epidemic.

While looking out for the mother, the breast is also looking out for the baby. It is constantly sussing out his or her nutritional and immunological needs. When the breast senses an infection brewing in the baby, it somehow tips off the mother's immune system and in turn the milk puts out more lactoferrin and the relevant antibodies. When the baby is older than one year, the milk contains more fat and cholesterol to match the baby's energy needs. When a baby is born prematurely, the mother's milk, as if anticipating its role, contains more protein and caloric density for a tiny tummy. Is it a coincidence or did we evolve that way, despite the unlikelihood that many preemies survived in our early history?

The breast is like a smartphone and juice bar in one. It communicates with the mother's body, the baby's body, and the environment. The breast knows the condition of the mother. Stress, for example, can cause her to hold back her output of milk. It can also send more cortisol into the milk, which appears to affect the long-term personality of sons (but not necessarily daughters), perhaps making them more exploratory or hypervigilant to grow up in a difficult environment. We know that a tough environment can affect a mother's stress levels. After the terrorist attacks of 9/11, many new mothers all over the country experienced temporary problems producing enough breast milk. My son was eight weeks old. Looking back, I wonder if this event played a role in our early difficulties with supply and demand. Unlike my mother, though, I never kept a nursing log.

Cells in the breast communicate with cells in the bone, telling the bone how much calcium to release for milk production and when to start guarding it again. A mother loses up to 6 percent of her calcium for her baby, but the stock more than fully replenishes within a few months after weaning. In terms of things like energy and minerals, breast-feeding takes a severe toll on mothers, though not as severe as the toll of gestating and delivering a fetus. Breast-feeding actually helps the mother recover from these events by tweaking her metabolism and protecting her heart. It's a critical part of how the enterprise was designed for our benefit as well as our baby's.

Internist and researcher Eleanor Schwarz had told a story in Peru about her inspiration for a study. "I was storing some of my milk in bottles in the refrigerator, and I noticed it looked like buttermilk," she'd recalled. "That's how fatty it is. Was there some relationship between the bottle of butter I was storing in the fridge and my future cardiovascular risk?" In other words, was the fact that her body was mobilizing her fat and siphoning it into milk helpful to her arteries or not? Some studies had shown that women who breast-feed lose more pregnancy weight than mothers who don't, but the data were inconsistent, and there wasn't much information about types of fat they lose or where it came from in their bodies.

Schwarz started crunching the numbers from the giant data set of the Women's Health Initiative, a long-term national health study. She found that while there wasn't much difference in weight loss between mothers who did and didn't breast-feed, women who didn't were 10 percent more likely to suffer cardiovascular disease and type 2 diabetes later on. She examined numbers from another study of 100,000 slightly younger women and found that those who breast-fed for just three months had a threefold lower risk of aortic or arterial calcification after adjusting for other lifestyle and economic factors. While pregnancy itself had put these women at risk for weight gain compared to women with no children, lactation had returned their lipid profiles back to baseline, a benefit that remained for decades. Mothers who nursed their babies also had less belly fat, which is known to be linked to heart and metabolic problems. "My current thinking," Schwarz concluded, "is that humans are mammals, and never lactating is not normal. I think breast-feeding plays an important role in a woman's recovery from pregnancy. It's like liposuction in terms of how much fat it's pulling off the body."

A nursing mother, of course, has no idea all this crosstalk is going on. She just knows her baby is hungry, and she can help. But food is a powerful currency. Add some hormones and she's a goner. Within hours of birth, my son and I (and then my daughter and I) became what scientists call "the mother-infant dyad" — a fully self-contained unit. For us, the hinge of it was a pair of breasts. Knowing that I could give my babies all they needed was nothing short of astonishing. Through breast-feeding, I grew more confident in my ability to be a mother.

Unfortunately, not many mothers get this

far. For such a clever, highly evolved system, it's too bad that breast-feeding is so ridiculously hard to do. I recently chatted with a young mother at a family gathering. Her eight-week-old baby girl was swaddled in pink, happily sleeping in her arms. "I tried breast-feeding for five days," said the woman, "but then the pain started. I was like, "that's it, I'm done." She's not alone; approximately 80 percent of newly lactating mothers have sore nipples, and many of them quit at the first sign of discomfort.

The big contradiction is that breast-feeding is so natural, and yet so completely unintuitive. What's really natural is for women to have a love-hate relationship with it, and this is something the lechistas don't tend to admit. Just as we've evolved to breast-feed, we've also evolved to be flexible, even whimsical, in our feeding habits. Some cultures and individuals nurse their children for years; others, not at all. In fact, humans are the only primates who wean their young long before they can forage on their own. We do this because we can, not because it's always the best thing for the baby.

Anthropologist Dan Sellen from the University of Toronto, who was also in Peru, said that most humans in foraging cultures weaned their young at about thirty months, "a pattern known to be optimal for growth and development." But there were always outliers — societies where the norm was nursing much further out or for much shorter periods. This flexibility — born from our human opportunism — may be natural, but it may also be problematic at the extremes enabled by manufactured formula, according to Sellen. As he pointed out, early weaning is sometimes okay from a nutritional perspective, but less so from an immunological one, especially as you look globally. When I returned from Peru, he sent me a recent paper of his. Its conclusion: the "mismatch between optimal and actual infant feeding practices in contemporary populations is widespread and presents a major public health challenge."

In a book about the unnatural (as well as the natural) history of the breasts, here is where we must deliver some more sobering as well as some strange news: not only is breast-feeding hardly practiced these days the ways nature intended, but the very stuff itself is oddly compromised.

In a great paradox of modern life, just as we're on the brink of truly understanding what's in human milk that can help us, the components are shifting.

As of late, breast milk has an unanticipated, new, ingredients label.

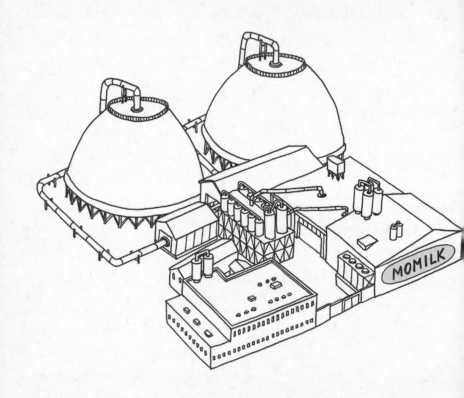

10
SOUR MILK

To recognize milk which is bad: thou shalt perceive that its smell is like *snj* of fish. To recognize milk which is good: its smell is like powder of manna.

— EBERS PAPYRUS,
ca. 1550 BC

If human breast milk, nature's perfect food, came stamped with an ingredients label, it would read something like this: 4 percent fat, vitamins A, C, E, and K, sugars, essential minerals, proteins, enzymes, and antibodies. It contains 100 percent of the recommended daily allowance of virtually everything a baby needs to grow, plus, as we've seen, a solid hedge of extras to help ward off a lifetime of diseases ranging from diabetes to cancer. Despite exhaustion, visiting relatives, and dirty laundry, every time we nurse our babies, the love hormone oxytocin courses out of our pituitaries like a warm bath. Human milk is like ice cream, penicillin, and the drug ecstasy all wrapped up in two pretty packages.

But read down the label a little farther, and the fine print sounds considerably less appetizing: DDT, PCBs, trichloroethylene, perchlorate, dibenzofurans, mercury, lead, benzene, arsenic. When we nurse our babies, we feed them not only the fats and sugars that fire their immune systems, cellular metabolisms, and cerebral synapses. We also feed them, in albeit miniscule amounts, paint thinners, dry-cleaning fluids, wood preservatives, toilet deodorizers, cosmetic additives, gasoline by-products, rocket fuel, termite poisons, fungicides, and flame-retardants.

If, as Cicero said, your face tells the story of your mind, your breast milk tells the decades-old story of your diet, your neighborhood, and increasingly, your household décor. Remember that old college futon? It's there. That cool paint in your bathroom? There. The chemical cloud your landlord used to kill cockroaches? Yup. Ditto, the mercury in last week's sushi, the benzene from your gas station, the perfluorooctanoic acid (an anti-grease coating) from your latte cup and sofa upholstery, the preservative parabens from your face cream, the chromium from your nearby smoke stack. One of the ironic properties of breast milk is that its high fat and protein contents attract heavy metals and other contaminants. If human milk were sold at the local Piggly Wiggly, it would exceed the federal safety levels for some of those chemicals in food.

On a body weight basis, the dietary doses our babies get are much higher than the doses we get. This is not only because they're smaller, but also because their food — our milk — contains more concentrated contaminants than our food. It's the law of the food chain, and it's called biomagnification.

To refresh that lesson from seventh grade, here's how it works: Animals at the top of the food chain receive the concentrated energy and persistent chemicals of all of the biota underneath them. Each member up the food chain takes in approximately 10 to 100 times the load of fat-loving toxins of its counterpart below. This is why a slab of shark meat contains more mercury than its weight in plankton. Ocean food chains are longer than terrestrial ones, so people who eat many marine carnivores carry higher body concentrations of some chemicals than the vegan who lurks at your local salad bar or even the steak lover next door. The Inuit, although they live in the remote Arctic with little nearby industry, are the most contaminated population on earth, besides victims of industrial accidents. But don't picture Eskimo Man in sealskin on the top of the food chain. Picture his suckling baby, who occupies yet another trophic level higher up.

If that's not creepy enough, some of the chemicals we pass on to our daughters will stay in their bodies long enough for them to bequeath them to their offspring. Even if we cleaned up our planet tomorrow, the industrial detritus of the last century has created a three-generation problem.

When I was nursing Annabel, I'd heard about this sorry state of affairs. I wanted to tell the story through what I learned from my own breast milk. There was a lot of emerging research about flame-retardants, in particular, because they were both so prevalent and believed to be disrupting hormonal systems in lab animals. At the time, only about a hundred other American women had had their milk tested for these substances as part of research studies, so I called Arnold Schecter at the University of Texas School of Public Health in Dallas. A medical doctor and professor, he'd been working for years on Agent Orange exposure and was now the go-to guy on flame-retardants. I would be Number 101.

I knew some of these chemicals would turn up in my breast milk; they are found in virtually every animal and human tested around the globe. The levels in American women are considerably higher than in anyone else, and they may be high enough to affect our health and that of our children. These levels tell us that our world is full of unhappy and improbable surprises, like the fact that our computer casings are somehow ending up inside of our breasts. Our collective levels tell us we cannot hide from toxins, no matter how carefully we shop, eat, and vacuum.

I felt pretty confident my levels would be relatively low. There was no basis for this assumption other than that I tend to live a healthy lifestyle. I don't smoke; I exercise. Since I had been pregnant or nursing two children for almost four years, I had been buying mostly organic food. Several years earlier we'd installed a three-stage reverse osmosis filter on our tap water and ice-maker. At the time, I lived in a leafy, scenic town in Montana. It was far from brown clouds and belching diesel freeways, although we did live near two Superfund sites, which are more prevalent in Montana than national parks are.

There was one obvious strike against me, which is that I was an older mother and therefore had stored up more long-lasting toxins in my body than a younger mother. In my favor was that I was nursing my second child, which means somewhere around half of my lifelong store of chemicals had already been siphoned through breast-feeding into my firstborn, then three years old. Nursing a baby, ironically, is the ultimate detox diet. As for my son, unless he learns how to lactate someday, he'll be stuck with my cast-off chemicals for a good long while.

Arnold Schecter had me ship my frozen vials of breast milk to Germany, where lab workers were capable of divining low levels of contaminants through sensitive mass spectrometer chromatographs. Schecter would interpret the results in Texas. Waiting for results over the ensuing two months, I learned more about the pervasiveness of flame-retardants in everyday American life. I started to get a little more anxious. It was a creeping discomfort, akin to when a person next to you on an airplane hacks up a nasty cough, over and over. I eyed my son's adorable foam airplane chair and the small tear in the car upholstery. I watched my son. Was he meeting his development targets? How was his attention span? I recognized that in its incremental way, alarm over toxic contamination creates a perfect storm for the overanxious parent of millennials. Now in addition to worrying about getting them into the right schools, dirty bombs, and car-seat recalls, we get to wonder whether our loveseats are emitting developmental neurotoxins.

One thing became clear: we live in a flame-retardant nation.

To understand how the substances ended up in everyone's breast milk, one must go back to 1938, when a German chemist named Otto Bayer invented polyurethane to replace rubber from markets cut off by war. First a rigid polymer, it was used to coat warplanes and to line the soles of Nazi boots. By 1954, chemists could fill the carbon-based material with air bubbles, creating flexible foam. It was practically an industrial miracle: cheap, soft, and malleable. It transformed everything from refrigeration insulation to upholstered furniture to car bumpers and tires. As one industry website proclaimed, "Today, polyurethanes can be found in virtually everything we touch — our desks, chairs, cars, clothes, footwear, appliances, beds, the insulation in our walls, roof and moldings on our homes."

There's just one problem: it's highly flammable, earning nicknames like liquid gas and fatal foam. A typical home filled with polyurethane products can literally burst into flames in five minutes once the petrochemical gasses heat up enough. Much household and office foam is treated with flame-retardants designed to delay ignition. The substances, which include bromine, chlorine, and phosphorous versions, came into widespread use after 1975 — when California, under pressure from the bromine industry, passed a consumer safety standard for furniture flammability. Specifically, it required furniture to resist ignition for twelve seconds in tests with open flames and smoldering cigarettes.

It's questionable whether or not these substances actually save lives. Nationwide, deaths from house fires are down on the order of 50 percent since the 1970s, a fact mostly attributed to fewer smokers and more smoke alarms. In fact, fire deaths have declined at a lower rate in California, the state with the most flame-retardant furniture (and the world's highest recorded levels of these toxins in human tissue). Flame-retardants may buy you a few extra seconds before ignition, but the chemicals make the smoke, soot, and fire worse by releasing other by-products like dioxin, a known human carcinogen. Most deaths in fires are caused by inhaling smoke or toxic gases.

In order to reduce flammability, foams, plastics, and fabrics are soaked or coated in these mixtures, which have commercial names like Firemaster 550 and V6. In some furniture-foam pieces, these mixtures account for up to 30 percent of their weight. (A common misperception, though, is that mattresses contain these chemicals. They don't, in case that helps you sleep better.) Most of the health research has been focused on a group of flame-retardants called PBDEs, polybrominated diphenyl ethers. A class of organic compounds, PBDEs have as one of their signature properties lipophilia, or fat attraction — hence, their unwelcome appearance in our breast milk.

Breasts, it turns out, are a particularly fine mirror of our industrial lives. They accumulate more toxins than other organs and process them differently. We first clued into this in 1949, when a Westport, Connecticut, doctor named Morton Biskind examined a pregnant woman who had strange neuropsychiatric symptoms, including vomiting, muscle weakness, and "unbearable emotional turbulence." He had been following scores of patients with acute poisoning from exposure to the ubiquitous pesticide DDT, which had hit the U.S. market a few years earlier. He'd heard the substance was being found in the milk of cows, rats, and dogs, so after the woman gave birth, he thought to test her breast milk. It was rich in DDT. Two years later, the scientist E. P. Laug published a study finding DDT in the milk of African-American mothers in Washington, D.C. In 1966, a Swedish researcher tested his wife's breast milk for PCBs — polychlorinated biphenyls, used to insulate electrical transformers — after he discovered them in dead eagle tissue. Two years later, Sweden banned them, with the United States following in 1978.

Because of the widespread use and persistence of PCBs, they are still among the highest-concentration toxins found in breast milk, even from mothers born well after the ban. One of the few chemicals to be banned outright in the United States, PCBs have been well studied. In humans, at elevated levels they can interfere with thyroid functioning and contribute to such gender-bending problems as masculinized female infants and feminized male infants. Studies also show a relationship between exposure to PCBs and breast, liver, and gall bladder cancer and lymphoma. Researchers in the Great Lakes region, the Arctic, and the Netherlands found that babies born to mothers with mid- to upper-range levels of PCB contamination (all due to eating diets rich in fish) have delayed learning capabilities, lower IQs, and reduced immunity against infections. Some problems have persisted into early adolescence, so far.

Molecularly, the flame-retardant PBDEs are eerily similar to their older cousins. But we may never have noticed flame-retardants showing up in living cells if not for a mysterious bout of animal poisoning in Michigan in 1974. That year, farm animals began suddenly losing weight, aborting calves, not lactating properly, salivating excessively, and suffering from diarrhea. A sluggish investigation followed, during which an enterprising government chemist finally discovered that workers at a plant that processed both animal feed and flame-retardants had inadvertently mislabeled the bags going to market. It was a classic Wile E. Coyote mix-up. Some foam got treated with food filler, and millions of animals statewide ate a bunch of flame-retardant —

in this case, a compound called PBB. Eventually, ten thousand cows, two thousand pigs, four hundred sheep, and two million chickens had to be slaughtered, but not before nine million people ate contaminated dairy and meat over the course of a year.

In the years that followed, some adults who ate these products reported immune and thyroid diseases, acne, miscarriages, and other problems. A number of babies were also exposed to the substance in utero or through breast milk, and several EPA-funded studies have tracked their health outcomes. The girls who were exposed to the highest levels of PBB through breast milk reached puberty nearly a year earlier than those exposed only in utero or whose mothers nursed them with less contaminated milk. Exposed boys had higher rates of urogenital birth defects and slower development. The girls, now women, are showing higher rates of miscarriage.

The Michigan event did not make huge news nationally, but toxicologists took note. When the Michigan-made flame-retardants were quietly taken off the market, PBDEs stepped in as replacements. Today we know the new compounds were not much safer. Swedish researchers first found them in river pike in 1981. Unlike PCBs and DDT, whose levels were gradually declining world-wide, PBDE levels were steeply rising. The Swedes decided to look for the chemicals in stored human milk samples, and what they found rocked the scientific community: from the early 1970s, when they first appeared commercially, to 1998, PBDE levels in breast milk were doubling every five years.

"No one had ever heard of them. We thought it was just a European problem," said Kim Hooper, a toxicologist now retired from the California Department of Toxic Substances Control. "So we looked in San Francisco Bay seal blubber, and found a 100-fold increase over ten years." Then the Americans, too, decided to look in human milk. When European scientists first saw the test results of American women, they thought there must be a mistake. Our levels were ten to forty times higher — a full order of magnitude — than those in women in Europe. The numbers were also doubling every five years, a rate unmatched by any known chemical since the 1960s. In the past two decades, our levels have risen so sharply that the curve resembles a rocket launch.

The most common flame-retardant found in human breast milk is a mixture known as "penta-BDE," which was until 2004 manufactured by the Great Lakes Chemical Corporation based in West Lafayette, Indiana. The company was making over twenty million pounds of it per year, most of it to drench flexible polyurethane foam. In 2004, however, the European Union banned this substance and another called "octa-BDE" (the names refer to how many bromine atoms bond to carbon rings in the molecule), concluding they were likely to be "chronically toxic in humans." For its part, the U.S. government instituted a voluntary phase-out in production but allowed foam manufacturers to continue to use large existing stockpiles. Another compound, "deca-BDE," is set to be phased out beginning in 2013, but a major replacement, deca-ethane, is almost identical in structure. Most of us will continue to be sitting on and ingesting PBDEs and their chemical cousins for a long, long time.

When Arnold Schecter called with my PBDE results, he had mixed news. The "good" news in relative terms was that at 36 parts per billion, my levels were only two points above the U.S. median. This means that roughly half of women tested had levels above mine and half below. The bad news was that my levels were presumably higher before I nursed my first child, and they were still many times higher than those of the rest of the industrialized world. What this meant, though, in absolute terms, remained unclear.

We knew that in rats, PBDEs bind to thyroid transport proteins and interfere with proper brain growth. "The most sensitive health endpoint is the harm done during development of the fetus and child," said Tom McDonald, then of the California Office of Environmental Health Hazard Assessment. "We've seen irreversible and permanent behavior in the learning abilities and memory in rats and mice. At low doses, quite frankly, we've seen effects to the reproductive organs and delays in puberty. If you look at tissue concentrations, the levels [of PBDEs] in some people are comparable right now."

Based on the limited bio-monitoring tests done so far, it appears that 5 percent of the general U.S. population may have unusually high levels of flame-retardants in their bodies for reasons nobody understands. Their levels, over 400 parts per billion, correlate to the levels at which harmful effects are seen in lab animals. Several recent animal studies indicate that PCBs and PBDEs may act in unison to block protein receptors and affect thyroid and endocrine functioning.

When I first started writing about PBDEs six years ago, we knew little about their health effects in humans. But since then, the compounds have been better studied, and the results have not been reassuring. In humans as well as rodents, they are believed to interfere with thyroid hormones. These, in turn, play a strong role in brain development and the regulation of metabolism, ovulation, the menstrual cycle, and fertility. You don't want to monkey around with a nursing baby's thyroid. In 2010, researchers in New York found that babies and toddlers with the highest PBDE levels scored lower on tests of mental and physical development, including verbal and performance IQ.

In adults, low thyroid levels — which I happen to have — have been linked to breast cancer. A recent study found that California women were 30 percent less likely to become pregnant in a given month for every tenfold increase in blood PBDE level. The authors concluded, "Because exposure to PBDEs is ubiquitous in industrialized nations, even small decreases in fecundability may have wide-reaching public health impacts." Males may be affected as well. A Danish study showed an association between PBDE levels and the incidence of male genital birth defects and lower birth weights. None of this bodes well for infant exposures.

To learn more about the recent science, I attended an international conference in San Antonio, Texas, called simply "Dioxin." The weeklong meeting, which convenes annually, is devoted to sharing the latest data on POPs, or persistent organic pollutants, including but not limited to halogenated flame-retardants. All POPs have as their defining characteristics a long life, widespread distribution, and well-established toxicity. Today we all live in the same chemical ocean with better and worse eddies. Not only do many synthetic substances accumulate up the food chain, but also they move great distances, carried by rain, wind, snow, and ocean currents. Lest you doubt these chemicals are everywhere, the conference sessions included such titles as "Flame Retardants in the Serum of Pet Dogs," "Evaluation of Human Toenail as a Noninvasive Biomonitoring Matrix for Assessing Human Exposure to Environmental Organic Pollutants," and "Human Exposure to Fluorinated Ski Wax."

After a mariachi band played a welcome, I cornered Linda Birnbaum, the director of the National Toxicology Program at the National Institute of Environmental Health Sciences. Birnbaum is a bureaucrat, but she's respected for her forthright views. I asked her if she believes people are harmed by flame-retardants. "Because we are all exposed on a continual basis, they will cause problems for some people," she said. "We all have different vulnerabilities: age, genetics, pesticide exposure, co-exposures, or cross-exposures. Our old way of looking at these things is one chemical at a time and that doesn't protect us." We are the sum of our chemicals.

If POPs are so bad for us, wouldn't we already be seeing a rise in the number of people with related health effects? Birnbaum said we might be seeing exactly that. She cited a rising prevalence of thyroid disorders, infertility, and learning and behavior problems. Although we don't know exactly what has triggered these problems, Birnbaum says chemicals like halogenated flame-retardants "will eventually build up to levels of concern, and we need to monitor and reduce them," she said.

She does not go so far as to suggest women stop breast-feeding. But since we'll be living with these substances for a good long while, it's certainly a reasonable question for women in industrialized countries to consider. To answer it, it's important to have some perspective. Breast milk is, after all, just one source of these chemicals for the baby. If they exist in milk, they exist in blood, and the infant's first and arguably most significant exposure takes place across the placenta during fetal development. Another huge exposure takes place in the home, just by lollygagging around, crawling on the carpets, sucking on fingers, and orally exploring the (flame-retarded) world. My kids thought cell phones and remote controls were pacifiers. Although it is known that breast-fed infants and toddlers have considerably higher levels of the chemicals in their bodies, their formula-fed peers catch up by mid-childhood.

Then there is the argument that breast-feeding may actually protect infants from the effects of chemicals, even as it is exposing them. Some studies have found that breast-fed babies develop better despite the additional chemicals found in breast milk, which is why the World Health Organization and other groups continue to recommend breast-feeding even among the Inuit, whose breast milk could technically qualify as hazardous waste.

Some lactivists are reluctant to highlight

breast-milk contamination because they don't want women to have yet another excuse not to breast-feed. This may be a real concern, but in a long history of such behavior it comes across as yet another condescending way to treat pregnant or nursing women. We'll tell you what you need to know! Trust us! When a California state senator proposed a statewide breast milk bio-monitoring program a few years back, an activist group opposed it, citing fear among mothers.

Yet breast milk is both a real reflection of our body burdens and a powerful symbol of contamination. "We test breast milk because it is a big sample with a lot of fat, it's chemical-rich, and it's noninvasive," California's Hooper told me. "We should be searching for fetal contaminants, and breast milk is the easiest way to look for them. It's a direct reading of what the fetus is getting." And let's face it, breast milk carries some political weight. It was not until persistent organic chemicals began appearing in human milk that countries took steps to ban them. "Clearly breast milk speaks louder than sediment," Hooper said. "When breast milk speaks, people listen."

But despite the reassuring arguments to just keep breast-feeding, I find myself unsettled. The amount of chemicals that infants suckle through milk isn't insignificant. Recent studies show that lactating mothers off-load about 2 to 3 percent of their total PBDE body burden per month to their offspring, or about 30 percent if they nurse for a year. I nursed both of my kids for eighteen months, and now I can't help but wonder if that was such a great idea. For other chemicals, the dump rate is even higher, with a range of up to 14 percent per month for dioxins and up to 8 percent per month for PCBs. (I know I had these substances in my milk as well, because we tested them for good measure. Sorry, kids!) Mothers who breast-feed for a year also siphon off to their infants 90 percent of their body burden of perfluorinated compounds, known as PFCs. Used in the manufacture of products such as Scotchgard, GORE-TEX, and Teflon, PFCs have spread across the globe, even ending up in polar bear tissues, and they virtually never break down in the environment. An EPA panel concluded one type of PFC called PFOA was "likely to be carcinogenic in humans."

Lately, the industrial world's enthusiasm for breast-feeding is being somewhat and very carefully modulated. As I learned at the Dioxin conference, the Norwegian

Scientific Committee for Food Safety is currently debating that country's breast-feeding recommendations, a fact that would have been unthinkable a decade ago. "I don't think it will change the current recommendations, but maybe there's no benefit to breast-feeding after six months," Cathrine Thomsen of the Norwegian Institutes of Public Health told me. Consider the gravity of this statement. Norway has the single highest breast-feeding rate in the world, with 99 percent of new mothers doing it. At six months, more than half of all babies are still nursing. The country has actually banned advertising by formula companies. It grants paid maternity leave for forty-two weeks. This is a country deeply committed to breast-feeding. And now it is rethinking it.

In Sweden, as Åke Bergman, a respected researcher there told me, "there is now a strong recommendation that women not lose weight during nursing," because that further mobilizes the contaminants in her fat and sends them into her milk. "It is like upending the candy bowl," he said, with the brightly colored candy being the POPs nestled in and among our fat cells. If you're scratching your head because you'd heard that losing weight was part of the point of breast-feeding, you're not alone. The old rules no longer apply.

Thankfully, our overall levels of these compounds are still quite low. But one can almost imagine a grim sci-fi future in which firstborns are sacrificed or discarded like bad first pancakes. Incredibly, this may already be happening in marine mammals. Adult female striped and bottlenose dolphins are actually the "purest" of all, because they have so effectively dumped (the technical word is *depurated*) up to 91 percent of their chemicals into their offspring, especially the firstborn. It's a much greater transfer than what occurs during gestation. And because their milk is so much fattier than ours, it also contains a much higher concentration of pollutants such as DDT and PCBs. The levels in firstborn calves sit above the threshold for which serious health effects can be expected, said biologist Randall Wells, a senior conservation scientist with the Chicago Zoological Society. And in fact, firstborn bottlenose calves off the coast of Sarasota, Florida, experience a much higher mortality rate than their younger siblings, about 70 percent versus 40 percent. This could be due to a variety of factors (such as maternal inexperience), but Wells said he wouldn't be surprised if pollution plays a role. "It does make me concerned for mammals," he said. "I wish we didn't have to contend with pollution along with everything else."

If we want to reduce our to exposure to these compounds, we need to know more about where they come from. Wanting to trace how those flame-retardants in my breast milk got there, I decided to start with my house dust. Because the compounds are not molecularly bound to the foam, they easily migrate out of their products and attach to dust, where they can be inhaled or ingested by women like me. So I collected dust from my vacuum cleaner bag and sent it off to Duke University. There toils Heather Stapleton. A dedicated young environmental chemist and new mother, Stapleton has become something of a flame-retardant queen. She recently made a splash by figuring out that house dust, not food, is our major source of exposure to PBDEs. (Of course, breast-feeding infants are an exception to this rule, since they do get most of their load from milk.)

A couple of months later, Stapleton sent me a spreadsheet. Each "congener" or molecular type of PBDE (there are dozens) leaves a fingerprint, and they can be traced.

Once again, my self-image as a granola girl was dinged: I had average-to-high levels of these substances in my home. For example, one congener, deca-209, is found in the hard polystyrene backsides of TVs and monitors. This is the flame-retardant not yet phased out in the United States. Based on a study of homes in Boston, the mean level for this congener in house dust is 4,502 parts per billion. My dust's deca-209 level was 5,279, probably because we have two home offices. I shouldn't feel so bad, though; one house had 185,600. What were they doing in there? Was that Mark Zuckerberg's dorm? On the other hand, the mean level for penta-47 (used in foam furniture) is 1,865. My level was 632. My level of octa-203 (used in electronics) was nearly triple the mean of 3.6. The portrait of my home, in other words, is pretty typically plushed, wired, and, well, dusty.

On the upside, dust studies of homes in the United States and Europe show that PBDE levels are starting to drop modestly, reflective of the bans and phase-outs. Our levels in milk will soon be leveling off and dropping as well. But hold the champagne. Flame-retardants are like the story of Hydra. You cut off one head and you get eight more. Since PBDEs have waned, some seventy-six new and suspect flame-retardants have taken up their call. My detective work was not yet complete.

Last year, I purchased an inexpensive couch from IKEA for the basement. One of the reasons I turned to IKEA is because, I, flame-retardant reporter, had heard IKEA's claim that it was no longer using brominated compounds. I was feeling smug. Then Stapleton invited me to send her a small chunk from several foam items for a furniture biopsy. (You know we're living in a strange world when we have to biopsy our furniture.) I cleaved off one-inch cubes from the couch and several random cushions I had around the house and shipped them to Duke.

The cushions came back clean, but Stapleton found that my couch contains a flame-retardant called 1,3-dichloroisopropyl phosphate, better known as "tris," or TDCPP. This was one of two notorious flame-retardants used in children's pajamas in the 1970s, until scientists linked the chemicals to DNA mutations. Public outcry forced clothing manufacturers to stop using them. Most of the research at that time was done on brominated tris, while the type in my couch is chlorinated tris. The theory goes that the chlorinated version breaks down more easily in the environment but is more volatile, meaning that more of it escapes the foam. It probably acts in similar ways to brominated tris in the human body. The U.S. Consumer Product Safety Commission classifies TDCPP as a probable human carcinogen, and the EPA considers it a moderate cancer hazard. In 2011, California added tris to its list of carcinogens under Proposition 65.

The first time Stapleton's mass spectrometry machine identified tris, she couldn't believe it. "At first, I thought, no one could be using this after all the concern in the '70s. We were shocked," she said. But recently, Stapleton has found tris in a lot of IKEA products as well as those made by other companies, including thirty-three of one hundred tested baby products such as nursing pillows and car seats.

IKEA, to be fair, has few options. It does not put these substances in furniture sold in any countries other than the flame-retardant-loving United States and the United Kingdom. Interested in what IKEA had to say about all of this, I e-mailed Bjorn Frithiof, a chemical specialist in the company's laws-and-standards department in Sweden. He responded as soon as the sun rose over Småland. He seemed both matter-of-fact and contrite. "IKEA aims to refrain from the use of chemicals and substances that could potentially be harmful to people and the environment," Frithiof wrote. "IKEA is currently phasing out all flame retardants of the chlorinated tris type from our products. This work is well on its way." In a later e-mail, he said the company would replace tris with "an organophosphorous compound which gets incorporated into the polymer matrix of the foam filling. It is a bit early to say if this solution will be the dominant one for our products."

There's nothing great about organophosphates, which can also persist in the environment. But if IKEA can figure out how to bind the chemicals to the foam, that would help. It would mean the stuff wouldn't waft out and work its way into our breast tissues every time we plunk down on the couch.

As my flame-retardant adventures made clear, there's only so much an informed consumer can do. A better solution would be a regulatory one. California should get rid of its outdated flammability standards. Congress needs to update its chemical laws so these substances can be tested for health effects before they come to market. Many scientists and activists and even some regulators advocate taking a precautionary approach to chemicals that exhibit the big trifecta of concern: persistence, toxicity, and easy transportability.

"If you know it's persistent, bio-accumulative, and toxic, it will be a problem," Linda Birnbaum of the National Institute of Environmental Health Sciences had stated flatly. "Since we know, why would we use them? Do we really need these chemicals or are there alternatives?" she'd asked. "The problem," she'd continued, "is the term *precaution* has gotten a bad rap internationally. People think it means uncertainty. There's always uncertainty, but there's also information. What we should do is act in the presence of information and not require certainty. It's not the same thing as acting in absence of all information."

But don't hold your breath for regulatory changes anytime soon. Several bills to reform California's flammability standard have failed, much to the disappointment of Arlene Blum, a chemist who first published her findings on the hazards of pajama tris more than thirty years ago and who now runs the Berkeley-based Green Policy Science Institute. Blum led the first American expedition to climb Annapurna and walked two thousand miles across the Himalayas;

she said both were easier than fighting the chemical industry.

Breast-feeding is an ecological act, connecting our bodies to the world in a complex web of give-and-take. The permeability of breasts allowed us to make great advances. Their estrogen sensitivity allowed us to reach puberty at optimal times. When our early ancestors migrated and settled in river and coastal areas, omega-3–rich diets turned their breast milk into gold, and our brains grew. We recruited, harvested, and bred specialized bacteria for our milk; we collected molecules from the world and from our bodies to manufacture novel sugar and fats to protect our babies. Our special low-protein milk kept us growing slowly, so we could have the longest childhoods on earth and learn everything we could.

Our brains grew so well that eventually we learned how to change the world's ecology. We couldn't possibly have guessed that we were changing our breast milk as well. Our *nouveau crème* no longer serves us as well as it once did. Ironically and tragically, as breast milk once propelled our evolution, now it may be impeding it by conveying toxins and quite possibly contributing to infertility and brain and body impairments.

For many decades, the formula companies have tried to mimic breast milk, but it is breast milk that now may be approximating formula. That is decidedly depressing.

In 2004, the United Nations implemented the Stockholm Convention on Persistent Organic Pollutants (POPs), whereby 162 countries agreed to ban or severely restrict twenty-one of the worst persistent organic pollutants, the world's most egregious chemical offenders. Most are pesticides, but several are PBDEs, along with PFCs, dioxin, and PCBs, all the top-billed breast-milk additives.

The United States has not ratified the treaty.

11
AN UNFAMILIAR WILDERNESS: PERIODS, THE PILL, AND HRT

Brave new world: A world or realm of radically transformed existence, especially one in which technological progress has both positive and negative results.

— *AMERICAN HERITAGE DICTIONARY*

Walking today in an unfamiliar biochemical wilderness, women's bodies are reacting unpredictably. Breast cancer may very well be one of those reactions.

— JAMES S. OLSON,
Bathsheba's Breast

In the course of a lifetime, breasts meet many friends and foes: lovers, babies, ill-fitting undergarments, persistent pollutants, maybe a nipple ring, a baggie of silicone, or a dose of therapeutic radiation. It's a lot to ask of breasts. Some of them don't make it to the finish line. Each year in the United States, some seventy-eight thousand women undergo mastectomies of either one or both breasts. Over ten years, that's enough women to fill the entire city of San Francisco. There are approximately two hundred thousand cases of breast cancer per year and forty thousand deaths in the United States. Globally, breast cancer is the leading cause of cancer-related death in women. One million women get diagnosed each year, and that number is expected to increase 20 percent by 2020. The global rise will largely be due to longer lifespans, obesity, and better screening. In a word, modernity.

Evolutionarily, though, cancer is a fact of life for multicellular organisms. As the pioneering German pathologist Rudolf Ludwig Karl Virchow discovered in 1855, *"omnis cellula e cellula,"* from one cell grow all cells. We need our cells to divide and replicate. Complex creatures like us would never have seen the sun if it weren't for the lucky trick of mutations.

Breast cancer has probably been around for as long as there have been breasts. Humans are just about the only free-ranging animal (other than minks) to get it. Domestic pets, if not spayed, get it. Breast tumors can be artificially induced in many other animals in the lab by feeding them strange carcinogens or implanting rogue cells or tissues or messing with their genes.

In some human populations, breast cancer is virtually unknown. Among the Kaingang women in Paraná, Brazil, there is reportedly no breast cancer. Researchers would love to know why. It's probably a combination of things: the Kaingang women have a shorter life expectancy, they bear many children whom they breast-feed, and they don't take oral contraceptives or hormone replacement therapy. Also, spending many hours outdoors, they probably get a lot of vitamin D. They probably don't wear deodorant or underwire bras, both long-held suspects in the breast cancer mystery, but ones that have been reliably exonerated.

Although breast cancer at the rates we know it today is a modern disease, ancient Egyptian doctors were familiar with it. One papyrus recommends applying a plaster made from cow's brain and wasp dung to tumors for four days. In the Middle Ages, breast cancer was a known and feared disease. The most advanced treatment at the time was the application of insect feces. Anne of Austria, the mother of King Louis XIV, famously suffered from breast cancer. She died in 1666 after caustic remedies of arsenic paste and a butcherous attempt at surgery (without anesthesia, of course). As barbaric as treatments seem today — slash, burn, and poison — we have it a whole lot better than the king's mum did.

Bernardino Ramazzini was a Renaissance-era doctor who sought to understand the causes of many diseases, not just their treatments. Born in Capri in 1633, shortly after the teachings of Galileo were banned, he is today considered the father of environmental health. In 1705 he published his life's work, *De Morbis Artificum Diatriba,* or the Diseases of the Workplace. It's a lively and, to modern eyes, very amusing work. One of his chapters is titled "Diseases of Cleaners of Privies and Cesspits." That one makes me appreciate the writer's life, until I come to "Diseases of Scribes and Notaries." In it, he writes, "All sedentary workers . . . suffer from the itch, are a bad colour, and in poor condition." That one made me get up and go for a hike.

Among the many things that interested Ramazzini were diseases of women, including midwives. In the archaic translation I have, most of the *s*'s look like *f*s, rendering such pronouncements as, "Nurfes . . . are likewife fubject to various Difeafes in the Courfe of fuckling." Old-language charms aside, Ramazzini astutely noticed that breast cancer was most common in nunneries. "You'll scarcely meet with a monastery that has not fresh instances of this cursed [or curfed] plague," he writes. He admits he does not know the reason, but he calls the breasts and womb the "Fountains of Letchery," and speculates that it is the "mighty lascivious" among the nuns who will get cancer due to suppression of their sexual drives. Well, he was wrong on the reason, but his observation that nuns were the most vulnerable women remains a critical piece in the modern understanding of this disease.

Women of God did not have a lot of fuckling or suckling going on. What did that mean for the lecherous organs?

Turns out it was a good question. Even W. H. Auden pondered cancer in 1937: "Childless women get it / And men when they retire; / It's as if there had to be some outlet / For their foiled creative fire." Eventually, epidemiologists realized it wasn't foiled creative fire as much as foiled reproduction that caused the trouble. It was childlessness, not celibacy per se, that increased cancer risk. It took centuries to figure out how hormones are made and used in the body, and in fact, we're still learning. In the nineteenth century, doctors knew that breast tumors were often larger and more aggressive in premenopausal women, and they also figured out that an individual woman's tumor was sometimes a different size before her period and after. By 1895, a Scottish physician named George Thomas Beatson had experimented with removing the ovaries of cows and pigs. He saw that doing so made their udders shrink, and he speculated that it might make breast tumors shrink as well.

Doubting the conventional wisdom that the nervous system controlled most bodily functions, Beatson presciently observed, "I am satisfied that in the ovary of the female and the testicle of the male we have organs that send out influences more subtle . . . and more mysterious than those emanating from the nervous system, but possibly much more potent than the latter for good or ill as regards the nutrition of the body."

Beatson didn't know those ovarian emanations were estrogen and progesterone. Nor did he know there were hormone receptors in breast tissue and in many tumors, but he witnessed the hormones in action. In the first woman whose ovaries he removed, her breast tumor retreated and she appeared cured. It seemed miraculous, and news of it made a huge splash. Unfortunately, Beatson also did not know that the body partly compensates for the loss of the ovaries by pumping out estrogen and progesterone-related hormones from elsewhere: fat tissues and the adrenal glands. Four years later, his patient's cancer returned with a vengeance and she died.

It was only the beginning of a long and disappointing legacy of miracle cures turning out not to be so. But Beatson, like Ramazzini before him, was onto something.

Malcolm Pike, whom we met in chapter 7 tinkering around with pregnancy hormones, is a rare academic hybrid. A mathematician born in South Africa in 1935, he first migrated to Britain in 1956, studying with the famed epidemiologist Richard Doll. Epidemiologists are people who use statistics to study disease, and Doll was their dean. He was the man who established that lung cancer was caused by smoking cigarettes, a conclusion that may seem obvious in retrospect, but was far from it at the time. Arriving in California in 1973, Pike soon fell in with a group of American epidemiologists. By this time, breast cancer rates in the United States were increasing between 1 and 2 percent a year, a trend that was both surprising and baffling. In 1940, the number of women coming down with invasive breast cancer on an annual basis was 59 out of

100,000. By 1960, even after adjustments for an aging population, it was 72, and by 1990, it was 105 and rising, to the point where the disease now strikes 129 women per 100,000, or 1 out of every 8 women who reach old age. Worldwide, a quarter of all malignancies are breast cancer. The race was on — and still is — to figure out how breasts became so imperiled so fast.

Interestingly, though, Japanese women developed the disease at rates six times lower than Americans. The epidemiologists wanted to know why. They knew it wasn't due to genes, because once Japanese women migrated to the United States, their daughters' cancer rates caught up. Was it a virus, or perhaps something else contagious about the American lifestyle or diet?

Pike was familiar with the work of Beatson, and he suspected clues might be found in the difference between American and Japanese women's reproductive lives. In 1980, he headed to Hiroshima for six months to study the archives of the Atomic Bomb Casualty Commission. In documenting minute details about survivors of the bomb, the commission's medical records collected excellent data on everything from a woman's age of menstruation to the number of children she had.

Pike and a colleague found that Japanese women in 1900 first got their periods at age sixteen and a half on average, meaning their ovaries kicked in more than two years later than those of their American counterparts. Their age at motherhood and the number of children they had were similar, but when Pike looked at weight, the data once again stood out. The average menopausal Japanese woman weighed just 100 pounds; the average American, 145. When Pike analyzed blood samples, he saw that rural Japanese women (and most were rural then) produced only three-fourths as much estrogen as American women.

"We wrote papers explaining that these things, menarche [age at puberty] and weight, might explain half the difference in breast cancer rates between the two countries," Pike told me during a visit to his office in the Topping Tower of the University of Southern California's Keck School of Medicine in east LA. As he saw it, American women were simply exposed to more estrogen and progesterone over their lives. Betrayed by their own ovaries and fat cells, these steroidal compounds were somehow causing more cancer.

As early as the 1930s, scientists knew that elevated estrogen levels triggered breast tumors in mice, but it had not yet been proved in humans. In the early days of hormone research, drug companies eyed it and other steroids greedily. Cortisone, for example, mimicking corticosteroids produced in the adrenal glands, was good for treating arthritis. If anyone suspected trouble with estrogen, they looked the other way as it was so promising commercially. Estrogen was known to promote bone density and soften the skin. It could perhaps prevent miscarriage.

Making these steroids in the lab was profitable but complicated. Commercially viable amounts of estrogen could be derived from huge quantities of pregnant horse urine and was first marketed in 1942. It's still a main ingredient in Premarin, a drug given to menopausal women (more on that later). The first commercially viable quantities of progesterone and testosterone were isolated from wild Mexican yams in the 1940s. Progesterone is a natural hormone made mostly by the corpus luteum in the ovaries, and later, if pregnancy occurs, by the developing placenta.

The word *hormone* comes from the Greek word meaning "to urge on." Like other hormones, progesterone travels with its commanding messages through the bloodstream. Among its many powers, it inhibits the release of additional eggs so that we get pregnant only once at a time. During our Pleistocene past, it wouldn't do for us to be gestating many more than one or two fetuses simultaneously. (This is in contrast to, say, tufted-ear marmosets, who often bear twins and triplets sired by different fathers. The upside to this clever system is that many males in marmoset-land tend to be stellar dads and upstanding citizens, each thinking all babies in the troop are his.)

Progesterone's uncanny ability to prevent ovulation caught the eye of mid-century drug companies seeking a better source of contraception than condoms, *coitus interruptus,* or the rhythm method. The latter technique requires timing sex to avoid a woman's fertile days. Although condoned by the Catholic Church, it was known as Vatican roulette for its failure rate. There was clearly a demand for reliable contraception. But medical progesterone had to be given by daily injection or it wasn't strong enough. A Big Pharma contest was on to find a way to make it potent enough to deliver by mouth.

Chemist Carl Djerassi, considered the father of the pill, describes the quest in his 1992 memoir, *The Pill, Pygmy Chimps and*

Degas' Horse. Born the son of a syphilis-specializing physician in Austria in 1923, he fled the Nazis to arrive in New Jersey as a teenager. By his mid-twenties, he was heading a lab in Mexico City for the pharmaceutical company Syntex. There, in 1951, he successfully fabricated a crystalline progesterone by rearranging some of the bonds of another synthetic hormone. "Not in our wildest dreams did we imagine that this substance would eventually become the active progestational ingredient of nearly half the oral contraceptives used worldwide," he wrote. It would become the pill known as Ortho-Novum.

G. D. Searle brought a similar product to the market first, in 1960, under the name Enovid. It was tested on rats, monkeys, and women in Puerto Rico (because many U.S. states banned birth control at the time). By 1970, ten million healthy American women swallowed a little magic pill every day. That's when Congress, swamped by complaints of the pill's side effects — everything from nausea to headaches to fatal blood clots — held a series of safety hearings. As the scientists were learning, the ways hormones worked in the body were far more complicated than they appeared. Drug companies had tweaked the original progesterone-only formulation to include estrogen to prevent "breakthrough" bleeding between periods, a discovery that had been made by accident. But hormone doses in the first combination pills had to be high to ensure protection from pregnancy.

In electrifying testimony, Roy Hertz, a physician and scientist with the National Institutes of Health, denounced the pill as an experimental drug and warned of carcinogenic effects. "Estrogen is to cancer what fertilizer is to the wheat crop," he declared. Sales immediately dropped 20 percent.

Even though new, differently dosed pills would continue to be very popular, Djerassi knew the days of fast and heady progress in the field of synthetic hormones were ending. In 1973, the famed organic chemist opened a fortune cookie in San Francisco that summed it up: "Your problems are too complicated for fortune cookies."

Around the same time as his trip to Hiroshima, Pike wondered if the pill could be contributing to rising breast cancer rates. He knew the pill's progesterone substantially decreased the risk of ovarian cancer because in preventing the ovaries from releasing eggs, it effectively stopped cellular division and growth. (In the middle of the menstrual cycle, the ovary literally rips open to send forth an egg and then must repair itself.) But progesterone did nothing to prevent cellular changes in the breast, and in fact, it was known to *cause* changes in the breast. For many years, estrogen was fingered as the primary culprit in breast cancer, because it was known to cause cancer of the uterus in lab studies and also to make human breast cancer cells grow faster in a Petri dish. But it turns out that progesterone is just as bad, and possibly worse, for stimulating cell growth and division.

Whenever a cell divides and replicates, it invites errors, or mutations. Enough mutations (you need a bunch) will send cancer on its inexorable course. Pike's theory was that the more menstrual cycles a woman has in her lifetime, the more her breasts are flooded with whiplashing hormones. But when a woman is pregnant or breastfeeding — as was normally the case for much of her reproductive life before industrialization, or unless she was a nun — her breast cells usually behaved in cancer-protective ways.

"The problem with women today is they reach puberty at twelve or thirteen and don't have a baby until they're thirty-five," said Pike. "That's extraordinary! It's incredibly non-evolutionary!" He's right; anthropologists have studied the cycling histories of contemporary hunter-gatherer cultures, believing they provide some insight into how early humans lived. Dogon women in Mali don't reach puberty until the age of sixteen, and soon after, they marry. They spend much of their adult lives either pregnant or nursing (they breast-feed each child an average of two years). They ovulate approximately one hundred times during their life. Women in Western nations ovulate, on average, four hundred times. Today in America, nearly 20 percent of women between the ages of forty and forty-four have never borne a child, a figure that has doubled just since 1976.

Pike scribbled down some charts and pictures for me on scrap paper and printed out articles on his printer. At seventy-four, he is tall and lean and generously bearded, like Santa after a crash diet. When I first Googled him, the top document was a marketing article from the University of Southern California calling him "the dashing Malcolm Pike," and it stuck in my mind. He speaks in a strong, lilting South African accent and enjoys asking Socratic questions. Having grown up in Johannesburg under apartheid, Pike is an arch proponent of tolerance and open debate. He clearly enjoys taking conventional wisdom apart.

"The pill," he continued enthusiastically, "gives you hormone levels *every day* that look like the levels after you've ovulated. It was a brilliant pathologist in Scotland who showed us that there's more cell proliferation in the breasts in the second half of the cycle."

"I could have told you breast cells are more active after ovulation," I said, thinking the genius in Scotland was a bit overrated.

The dashing Malcolm Pike raised his eyebrows. "How do you know?"

"Because they hurt! They get bigger. They're inflamed," I said.

"Ah!" said Pike. "But how do you know?" he repeated. "How do you know that's not just water? You can't feel cell proliferation! We had to see it in a dish!"

Well, I thought, *if that's the way scientists need to do it, fine.*

The reason breasts become bigger and more tender after ovulation illustrates how strongly these organs are geared to procreation. Every time an egg pops out of the ovary, the body is preparing for the big event, whether fertilization actually takes place or not. Progesterone courses through our cells to help prepare the uterine wall and to begin growing the dairy machinery in our breasts. It may seem overzealous to do this nine-plus months ahead of time, but in fact the breast needs every possible minute to get up to speed.

In any case, by the mid-1980s, Pike was publishing papers showing that women who began taking the pill as teens, before bearing children, doubled their risk of breast cancer before age forty-five. If they took the pill for eight years before becoming pregnant, they nearly tripled their risk. Captive tigers and lions also suffer from mammary and uterine cancers after taking oral contraceptives.

Between 1986 and 1989, a handful of studies in Europe and New Zealand confirmed Pike's human data, although other studies showed the pill added a smaller risk of breast cancer. I told Pike I took the pill starting when I was eighteen. By then, in the mid-1980s, manufacturers had introduced lower-dose pills, despite insisting all along that the original formulation was perfectly safe. Today's oral contraceptives contain one-fifth the hormone levels of the original.

"So has the pill transformed your breast?" he asked, anticipating my question. "We don't know. How would you ever find out?

You have to stick needles in people to look at breast tissue. We've been extremely reluctant. If we could look at 180,000 women, we'd understand."

Just when I was starting to feel the dread of past mistakes, he asked, "How long did you take the pill?"

I shrugged. "About four years," I said.

"We think the risk of breast cancer goes up for about ten years after you stop," he explained. "So there's probably no more risk for you. Now, whether it did you any good, we don't know."

Pike was untwisting a green paper clip, working it into a rough quadrangle. "We know it's protecting you from ovarian cancer."

I told him that after I stopped the pill, it took me six months to start ovulating again. He looked elated. "That four years you were on the pill was like having four babies when you were young!" he said. As far as my ovaries are concerned, that's a good thing.

But before breaking out the cigars, Pike turned again to breast cancer. If the pill gave him epidemiological heebie-jeebies, so did hormone replacement therapy, or HRT. Like the pill, this therapy supplied extra daily doses of menstrual hormones, estrogen and later estrogen-plus-progesterone, but to women whose ovaries were no longer making them. As early as 1982, Pike was worried about HRT. He published papers, but, as he tells it, they met with thunderous silence. By 1992, Premarin (the name stands for *preg*nant *mar*e ur*ine*) was one of the most widely prescribed drugs in America, given to 11 million menopausal women and earning its happy makers nearly $2 billion a year. To create the unprecedented demand, drug companies and physicians appealed to women's vanity and reason, essentially inventing a new pathology called menopause in the same way the surgeons had invented one called micromastia, for small breasts.

As one physician told the *New York Times* in 1975, "I think of the menopause as a deficiency disease, like diabetes. Most women develop some symptoms, whether they are aware of them or not, so I prescribe estrogens for virtually all menopausal women for an infinite period."

He had good company. In 1966, the prominent gynecologist Robert Wilson wrote an influential bestseller, *Feminine Forever,* in which he called menopause a state of "living decay" that makes women fat, moody, and saggy. He wrote that women "rich in estrogen," by contrast, "tend to have a certain mental vigor that gives them self-confidence, a sense of mastery over their destiny . . . and emotional self-control." Estrogen therapy, he wrote, "makes women adorable, even-tempered, and generally easy to live with."

Not unlike the anthropologists who believe that women's breasts exist for men, many mid-century doctors thought that women's moods, sexuality, and general perkiness should be engineered, artificially if need be, to suit male preferences.

By now there has been much debate about whether menopause is evolutionarily adaptive — in other words, is there something useful about it? — or whether we're really designed by nature to just wither up and die after our ovaries wear out. A common refrain is that we're more or less supposed to get cancer simply because we live so unnaturally long. I won't go much into the fray, but one of my favorite rejoinders, the "grandmother hypothesis," is well defended by anthropologist Sarah Blaffer Hrdy. In her book *Mothers and Others,* she explains that our ancestors often lived past the age of reproduction, and those grannies were in fact critical to the survival of their progeny for most of human history. Far from being a medical malady, menopause is a highly adaptive mechanism to free up older females to help feed and care for their grandchildren. No creature on the planet is more costly than the human child, who needs a ridiculous amount of time to grow up; without extra hands to supply the thirteen million calories needed until maturity, our species would not have made it this far. Women are supposed to outlive their ovaries after all, concludes Hrdy, with breasts happily intact.

Still, given the choice, who wouldn't want to have smooth skin and be adorable forever? Estrogen, miracle hormone that it is, does indeed relieve such symptoms as hot flashes, night sweats, and depression, which are experienced to a serious degree by 5 to 15 percent of menopausal women. This is the group that probably should have been targeted for risky drugs, but that wasn't nearly as profitable as targeting the entire sex.

When uterine cancer was linked to estrogen therapy in 1980, drug makers responded by adding progesterone to the formulations. Then numerous studies in the 1980s and 1990s linked other complications to HRT. In 1991, researchers launched the fifteen-year, $100 million Women's Health Initiative Study, but they abruptly halted part of it in 2002 when they discovered that the women taking HRT (as opposed to a placebo) had a 26 percent increase in the incidence of breast cancers, a 41 percent increase in the incidence of strokes, and a 29 percent increase in the incidence of heart attacks.

Britain was also conducting studies. Its Million Women Study, the largest study of HRT, yielded data in 2003. Results showed that women who were taking both estrogen and progesterone had a 100 percent increase, or a doubling, in their risk of breast cancer. The main culprit appeared to be progesterone, just as Pike had shown years earlier. Estrogen alone created a smaller breast cancer risk, and more recent studies suggest it might actually protect against breast cancer, but, alas, not against uterine cancer. Typical of the unpredictable ways hormones work their magic (and harm), HRT's added progesterone helped minimize one cancer (uterine) but exacerbated another — breast cancer. Overall, hormone therapy in Britain caused an additional twenty thousand cases of breast cancer during the decade of the study, which is still a relatively small added risk, but enough to give many women pause.

Although researchers had known for years

that synthetic hormones were linked to breast and other cancers, it really wasn't until these two huge studies in 2002 and 2003 that the knowledge finally stuck. To the dashing Malcolm Pike, that time lag was nothing short of tragedy. "People say, 'Aren't you proud? You spotted the danger early,' and I say, 'No, because our work didn't stop it,'" he said. "Why?" He shrugged. "Doctors are not particularly nervous, and they like prescribing things," he said.

But the answer also lies upstream, with the pharmaceutical industry, which has a zeal for profit and a masterful command of the regulatory landscape. These are traits it shares with the chemical industry. Both came of age in America at the same moment in time, often using very similar molecules.

It all adds up to a new ecology of the breast. Here's the basic cheat sheet for how the risk of breast cancer has changed over the ages. Old days: We did not have so much exposure to cycling estrogens and progesterones (we were thinner, hit puberty later, had more children, dropped dead earlier). Modern times: We're awash in steroidal hormones. We're fat and sexually precocious, but have babies late if it all. We take the pill and "bio-identicals," slather novel, untested chemicals on our bodies, and consume them through food and water. We're pretty much marinating in hormones and toxins.

Just as our long environmental legacy as synapsids, mammals, and primates shaped our cellular past, our modern environment — in the largest sense of the term — is determining our cellular destiny.

But you don't have to be in the sway of synthetic or natural hormones to get breast cancer. You don't even have to be a woman.

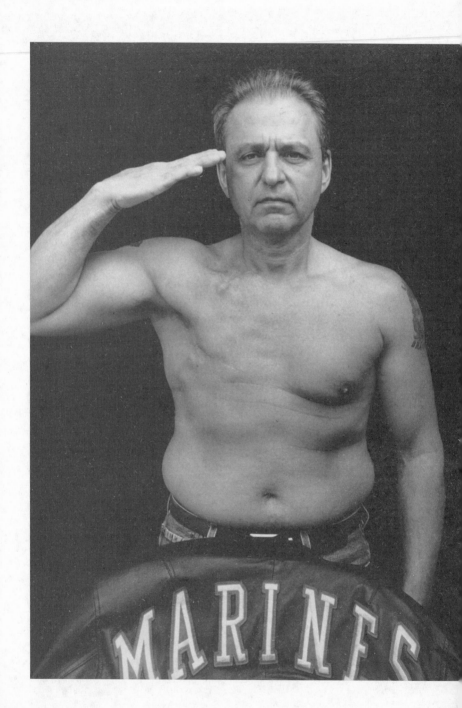

12
THE FEW. THE PROUD. THE AFFLICTED: CAN MARINES SOLVE THE PUZZLE OF BREAST CANCER?

Do unto those downstream as you would have those upstream do unto you.

— WENDELL BERRY,
Citizenship Papers

Even its name sounds jaunty: Camp Lejeune. Signs posted along the Marine Corps base in coastal North Carolina pointed the way to archery and bowling. The Burger King on Holcomb Boulevard advertised frozen fruit drinks. I half expected to see a pickup game of Capture the Flag. For a moment, I thought it looked like summer camp, until I realized it's the other way around. The traditional American summer camp model is based on the military. Think about it: the uniforms, mess hall, reveille, all those games of conquest.

People don't live at Camp Lejeune, they live "aboard" it. Home to some 150,000 marines and sailors and their families, the base covers 236 square miles. In its outer reaches lies evidence of the more serious pursuits of the Second Marine Corps Division. There is, for example, the (former) Live Hand Grenade Course, the Fortified Beach Assault Area, and, of course, the Flame Tank and Flame Thrower Range. (Don't think that one will be coming to Camp Wigwam any time soon.) These sites may have helped keep America strong, but they have also helped keep it sodden with volatile organic compounds. All of these zones, plus dozens of others on the base, are currently marked on the Environmental Protection Agency's National Priority List — otherwise known as Superfund.

Unfortunately, the base's worst historic contamination overlay much of its drinking water supply for at least three decades, from the mid-1950s to the mid-1980s. Nicknamed "camp sloppy," the base is made up of a big series of linked wetlands, aquifers, and the lazy New River flowing down the middle of it, all tilted toward the ocean. In one industrial area of the base known as Hadnot Point, fuel tanks silently dribbled or poured close to two million gallons of gasoline into the groundwater, forming a plume of petroleum now believed to be fifteen feet thick and half a mile wide. Atop it all sat well number 602, which in 1984 helped supply water to eight thousand people and yielded a reading of 380 parts per billion of benzene. This is seventy-six times the legal limit for benzene, a known human carcinogen.

Hadnot Point was known as a "fuel farm" — essentially the base's gasoline depot — and it's also where the Second Maintenance Battalion fixed tanks, jeeps, and other fleet vehicles. Beginning in the 1940s, it also held, in addition to gasoline, leaking storage tanks of industrial chlorinated solvents, notably trichloroethylene (TCE) and perchloroethylene (PCE) used for degreasing machinery. Up the road sat the base's disposal yard, where these solvents and others were dumped or buried. Some wells were more contaminated than others, but water from many wells was routinely mixed and then distributed to numerous houses and barracks from central water treatment plants.

The legal drinking water level for TCE and PCE, long considered probable carcinogens, is 5 parts per billion. That level, though, wasn't established until 1989. Although the military knew there was a potentially hazardous contamination problem by 1982, it did not routinely check the levels here until late in 1984. At that time, analysis from one well revealed 1,600 parts per billion of TCE. Tap water at the elementary school contained 1,184 parts per billion. That is five times the levels recorded in the poster-child-city of water pollution, Woburn, Massachusetts, site of the book and film *A Civil Action.*

Camp Lejeune, in addition to being the "home of the Marine expeditionary forces in readiness," now enjoys distinction as having had the most contaminated public drinking water supply ever discovered in the United States. Over the decades, 750,000 people drank it, bathed and swam in it, and inhaled its vapors.

The base also happens to form the center of the largest cluster of male breast cancers ever identified. We know this thanks not to the U.S. Marine Corps or even the Agency for Toxic Substances and Disease Registry (ATSDR), an arm of the Centers for Disease Control and Prevention that is tasked with assessing the health effects of the contamination. We know this because of one man with the disease and an Internet connection, Michael Partain. He calls himself Number One.

Partain, a father of four and an insurance claims adjuster in Tallahassee, Florida, was diagnosed with cancer in his left breast in 2007, at the age of thirty-nine. He underwent a partial mastectomy and eight rounds of chemotherapy, and then developed "gonadal failure," or an inability to produce testosterone. This was tough business for the son of a marine. "I never even knew men could get breast cancer," said Partain. "I kept thinking, what did I do to win this lottery? I never drank or smoked. I liked backpacking and Boy Scouts. There is no history of breast cancer in my family."

Not long after Partain was diagnosed, his father called him and told him to turn on the TV. There was a news report about the pollution at Camp Lejeune and its possible links to leukemia and other diseases. It was the first either man had heard of the contamination. Partain had been conceived and born on the base. "I knew right away I'd been exposed. I figured if this was the cause of my cancer, I wouldn't be the only one," he said. Partain went public with his diagnosis in the local media. Soon after, he got a call from a preacher in Alabama. He'd lived as a child in the same neighborhood as Partain, and at the same time. The preacher became Number Two.

Partain started Googling around for male breast cancer, and soon he found a photo of another man with the disease in Michigan. "His chest was half gone," recalled Partain, "and his Marines uniform was draped over his arm. I was like, *holy shit.*" Before long, he'd found twenty men with breast cancer and ties to Camp Lejeune. CNN ran a story on them and their conviction that their disease was linked to contaminated drinking water on the base. Overnight, twenty more men contacted Partain. Soon there were fifty.

As of this writing there are seventy-one of them, and the number goes up virtually every month. (There are also plenty of women around who lived on the base and have breast cancer, but what else is new?) Is there really a link between the men's cancers and the drinking water at Lejeune? Although some two hundred chemicals have been found to make mammary tumors grow in lab animals, it's been extremely difficult to link chemicals to the disease in humans. Many experts say there is only one proven environmental cause of breast cancer, and that's radiation. If new insight emerges from studying men like Partain, it could profoundly alter the way we view environmental health, and breast cancer in particular.

In the Western world, the incidence of breast cancer has grown in both men and women between 1 and 2 percent a year since 1960 (with the exception of a short-lived dip in women's rates in the last decade), although it is still very rare in men. For every one hundred women who get breast cancer, only one man does. But ironically, it may be the men who help solve the puzzle of this disease. In looking for a link between breast cancer and chemicals, it's much simpler to study men than women. Men's risk factors aren't complicated by such things as age at puberty, reproductive life history, and hormone replacement therapy. They're just guys with a very rare disease, and rare diseases are easier to trace to environmental exposures. This cluster, unlike so many others, could prove statistically significant.

"We stick out like a sore thumb," said Partain.

For all the personal tragedy Camp Lejeune may have caused an untold number of marines and their family members who have suffered from childhood cancers, birth defects, miscarriages, and adult diseases, the saga may prove a tremendous boon to scientists. Many of them gathered one recent summer day in Wilmington, North Carolina, for the twentieth meeting of the

Community Action Panel, a committee of experts and local activists put together by ATSDR. The panel, one of several centered on Superfund sites around the country, meets four times a year to discuss the state of the science and any concerns that community members have. Thanks to active public participants, researchers found out about the massive benzene plume. It's also because of the participants that male breast cancer, along with a number of other health problems, is now being studied by the agency.

Partain sits on the community panel, and so does a former drill sergeant named Jerry Ensminger, whose daughter, Janey, died of leukemia in 1985 at the age of nine. Two other panelists were notably absent, one suffering from battle-related post-traumatic stress disorder and the other recently dead of parathyroid cancer. Ensminger, bullish and compact, plays Calvin to Partain's more measured Hobbes. Ensminger and Partain were the two lead characters of a recent documentary titled *Semper Fi: Always Faithful,* about the former Lejeune residents and their battle against the military for truthful information and health-care benefits.

"You're wearing cowboy boots," Partain said to Ensminger as they walked across the local campus parking lot to the meeting.

"That's for kicking some ass," Ensminger replied.

After a round of introductory remarks, in which Ensminger lit into the U.S. Marine Corps for sending only a silent observer to the meeting, federal epidemiologist Perri Ruckart reviewed ATSDR's work to date. One health study performed by the agency in 1998 pointed to an association between the drinking water and male babies born smaller than expected. But in light of new (and more damning) evidence of contamination, the findings are now being reanalyzed with updated water modeling data. Several other important studies are ongoing, said Ruckart. These include a study of birth defects and childhood cancers in children born on the base, an overall mortality study of marines who lived there during the contamination, and a morbidity study that will canvass three hundred thousand former residents and base workers for illnesses. These studies will compare results with a similar but unexposed population from Camp Pendleton, a marine base in the state of Washington.

As Ruckart and her colleague Frank Bove had explained to me earlier, these are classic epidemiological studies, called case-control studies, which compare similar populations exposed to different things to see if one group is sicker. They're not perfect because researchers must rely on high percentages of people to enroll in the studies. If only sick people from Lejeune choose to answer questionnaires, that's called selection bias, and it can skew results. The ATSDR researchers won't have to rely on participants to tell the truth, as they'll laboriously confirm all medical diagnoses. These studies take years and cost tremendous amounts of money. The Lejeune studies will cost upward of $20 million. (Cleaning up the base is expected to cost more than $200 million.)

When they do work, health studies like these can be informative and (eventually) have a dramatic influence on public policy and medical practice. A case-control study is how researchers first reliably linked smoking to lung cancer. To date, the most revealing human cancer studies have been occupational ones, in which workers were known to be exposed to a particular chemical over a given length of time. Only a few general-population studies (as opposed to worker studies) have ever effectively proven a link between chemical exposures and cancer. Interestingly, though, two such studies also involved the solvents TCE and PCE, as well as other contaminants. In both Woburn (site of chemical and glue factories) and in Toms River, New Jersey (site of a dye-and-pigment plant), federal researchers concluded that the towns' high childhood leukemia rates were caused by contaminated drinking water, although they could not untangle one particular compound as the villain. In Woburn, an unusual number of male breast cancers also appeared, but the number was too small — only a handful — to be statistically meaningful.

That is why Ruckart and Bove are looking so hard at Camp Lejeune, where hundreds of thousands of men were exposed and already a high number with breast cancer have come forward. If nothing else, the numbers to work with are bigger, and that means more reliable. As ATSDR director Christopher Portier explained it to me after the meeting, "If I take a coin and flip it ten times and get seven heads, that could be biased by chance or not. But if I flip it one thousand times and get seven hundred heads, then I guarantee you there's an association. If there are real effects [to be seen at Camp Lejeune], then they will pop out in these studies where they looked marginal in others. We will do distinctly new science here."

At one point during the question-and-answer session, a local woman asked, "What can you tell this man here about the cause of his health problems?" Portier answered, simply, "Nothing." Even if the studies show a leak-proof link between cancer and the base's water, those conclusions would not apply to individuals, only to the risks faced by the population as a whole. Try telling that to Partain, though, who points out that the average age of onset for male breast cancer is seventy. "Over half the men I've identified are under fifty-six years old," he said. "That's not right. I know what caused my cancer."

Human nature is such that many of us easily believe causal links where they may not exist, especially when it comes to personal or familial tragedies. But the Camp Lejeune cluster has certainly raised the interest of academics and clinicians. Richard Clapp is an epidemiologist recently retired from Boston University who is serving as an outside expert on the community panel. He cautions that it may be years before the men get answers about the breast cancer. Even then the answers may be shrouded. While this is the largest cluster of male breast cancers ever found, there still might not be enough cases to get a strong signal in the data, he said. On the other hand, if an association does pan out, people will take notice. Most of Lejeune's pollutants are not known to act as hormones, "so it would make it more of a pure chemical story, and you could say at least one type of breast cancer can be caused by chemicals," said Clapp. "This should provide an opportunity to learn something. From an academic point of view, it's good. For the men involved, it's terrible."

Camp Lejeune may be a very troubled spot, but it's hardly unique. Now strongly suspected of causing human kidney cancer, TCE and PCE have been widely used both by the military and by many civilian industries. There are 130 military bases in the United States listed as National Priority sites under the Comprehensive Environmental Response, Compensation, and Liability Act, or the Superfund law. TCE alone has been detected in at least 852 Superfund sites across the United States, and the chemical is the most frequently detected organic solvent in groundwater. It is suspected of being present in 34 percent of the nation's drinking water supplies. TCE has primarily been used as a degreaser, septic-system cleaner, and dry-cleaning agent. In the what-were-they-thinking annals, TCE was also once a pet food additive, coffee de-caffeinating compound, wound disinfectant, and even an obstetrical anesthetic. The Food and Drug Administration banned these uses in 1977, but regulators did not formally limit TCE and PCE in drinking water until the late 1980s. In September 2011, the EPA formally reclassified TCE from a "probable" to a "known" human carcinogen based on solid evidence linking it to kidney cancer and suggestive evidence of neurotoxicity, immunotoxicity, develop-mental toxicity, and endocrine effects.

PCE, sometimes called "perc," is another chlorinated compound very similar to TCE. It is still used by most dry-cleaners, al-though the U.S. government is likely to stiffen regulations in the near future. Ben-zene, which is still a gasoline additive and smells vaguely sweet, was once used an af-tershave. Now, according to Bradley Flohr, assistant director for policy, compensation, and pension services of the U.S. Depart-ment of Veteran Affairs, "we know for certain benzene is associated with acute my-elocytic leukemia and other problems."

Unfortunately, there's more bad news:

both TCE and PCE degrade into a potent toxic molecule, vinyl chloride. That too has been detected at some of the well sites. Vinyl chloride was one of the first chemicals ever designated a known human carcinogen by the U.S. National Toxicology Program and the International Agency for Research on Cancer.

But very little is known about what, if any, role these compounds play in breast cancer. Several studies have looked at breast cancers in both male and female workers exposed to these substances, but the results have been contradictory so far, and the study sizes have generally been very small. One recent European study found a doubled risk of male breast cancer in motor vehicle mechanics. "Petrol, organic petroleum solvents or polycyclic aromatic hydrocarbons are suspect," it concluded. Vinyl chloride has been linked to breast cancer in workers making PVC plastic. Another study found a very moderately increased risk of breast cancer among aircraft maintenance workers exposed to TCE. Some studies found that dry-cleaning workers who are exposed to PCE have a higher incidence of breast cancer, but other studies found a lower incidence.

A 1999 study looking at Danish women

employed in other solvent-intensive industries found a doubled risk of breast cancer. Intriguingly, a set of studies looked at women on Cape Cod who had been unwittingly drinking public water laced with PCE from the lining in old pipes. Researcher Ann Aschengrau, an epidemiologist at Boston University's School of Public Health, found that the women exposed to the highest levels of PCE had a 60 percent greater risk of breast cancer than those who were less exposed.

Aschengrau points out that both TCE and PCE are fat-loving compounds known to accumulate in breast tissue at high concentrations. It's possible that special enzymes in breast tissue, notably in the ducts, might prevent the chemicals from breaking down. Once they're sitting there, they could plausibly damage DNA in rapidly dividing breast tissue. As we saw in chapter 3, men have breast tissue also, and some men have more than others.

Just as it may seem ironic to have men lead the way down this particular path in breast cancer research, it is now the U.S. military — long resistant to notions of environmental health — that stands poised to become a pioneer, albeit a reluctant one, of environmental medicine. Of course, the armed forces have a long history of making its uniformed ranks sick. There was ionizing radiation from the Second World War through the cold war, Agent Orange during Vietnam, and, most recently, burn pits spewing dioxin and other compounds in the Iraqi outback. If you look at the increasing compensation claims being paid out to former Lejeune residents, it appears that TCE, PCE, and benzene will soon join the list of culprits. Every step of the way, Congress had to prod the Department of Defense to study the sicknesses and offer fair redress.

For the VA doctors, it continues to be an education. Terry Walters, director of the Environmental Agents Service at the Veterans Administration, came to the Wilmington meeting and spoke with me during a break about the challenges to her agency. "Exposure issues aren't taught in med schools," she said. "But for physicians within the VA, that should be our stock and trade. We should be experts in this. Getting that education out to Podunk or wherever is a big, big challenge. It's not as mainstream as diabetes or cardiovascular disease. But every primary-care doctor should understand, if not specific information about benzene or TCE, then where to go to ask questions.

My hope is that when a veteran comes in and says, 'I was exposed to benzene,' he or she won't get a deer-in-the-headlights look from the doctor."

How likely is it that many of our friends' or relatives' breast cancers are caused by an environmental agent? It's virtually impossible to say. The American Cancer Society attributes only 2 to 6 percent of all cancers to chemical exposures, an estimate based partly on old and limited studies on occupational cancers. We learned about breast cancer and one environmental agent — radiation — from a large and unfortunate health experiment called the atomic bomb. For chemical exposures in a general population, though, confirmation is very difficult. We simply have too many mixed exposures over too long a time. "Epidemiology is what happens when you let all the rats out of the cages," joked Frank Bove. There probably never will be a simple "smoking gun" in the search for causes of breast cancer in the general population. Our world and our genes are engaged in too complex a dalliance for that. After all, there are numerous types of breast cancer, dozens of cellular and molecular pathways that can lead to them, and probably an untold number of factors, including genes, that can alter those pathways.

We do know that there are some hot spots for breast cancer near hazardous waste sites and industrial facilities. In Long Island, New York, Marin County, California, and Cape Cod, Massachusetts, breast cancer rates have risen faster than in the rest of the nation. These locales share a legacy of industrial, agricultural, and military pollution. But other factors are also higher there and confuse the picture: all are wealthy enclaves, where women have children later, take more hormone therapy, and drink more wine. No wonder the epidemiologists get their pants twisted up.

In the President's Cancer Panel report released in April 2010, the authors stated that cancers caused by chemicals have been "grossly underestimated." The authors of the two-hundred-page report took an unusually bold stance in urging better oversight of the chemical industry. Coauthor Margaret Kripke, an immunologist at the University of Texas MD Anderson Cancer Center, was once a skeptic on the topic of environmental disease. "I always assumed that before things were put on market, they would be tested," said Kripke. "I learned that is not the case. I was so naive."

Most of the major breast cancer organizations say there is no clear evidence that chemicals can cause breast cancer in humans. But in fact, there is little clear evidence that other things cause breast cancer, including the top favorites of obesity and smoking. If we look at all of the known red flags for breast cancer, such as reproductive and hormonal factors, family history, and radiation, they account for little over half of all breast cancers. Yet researchers have spent untold millions studying those things and very little studying chemicals. Perhaps it's time, say many activists, to look deeper into chemical exposures, especially since damning evidence in animals and in occupational studies is slowly mounting. The existing research is troubling enough that in 2010, the Susan G. Komen for the Cure foundation broke ranks to shell out $1.2 million to the National Academy of Sciences for a major review of the science on environmental exposures and breast cancer. The result: while chemical causes seem "plausible," better science is needed.

"A lot of data do suggest chemicals cause tumors in mammalian systems," ATSDR director Portier told me after the community meeting in North Carolina. He believes the environment, defined broadly to include smoking, nutrition, and chemical exposures, causes most cancers. "We have good studies now, for example with identical twins, that suggest the numbers could be as high as 75 percent, he said.

"There's a growing body of animal evidence and sporadic human evidence that things we're exposed to across a lifetime can cause breast cancer," agreed Marion Kavanaugh-Lynch, director of the California Breast Cancer Research Program, which funds environmental studies with proceeds from the state's cigarette tax. "If we can identify these chemicals now, we can more easily avoid them," she said.

It's too late for number twenty-three on Partain's spreadsheet, who as an infant on the base attended a day-care center in the early 1980s that had been converted from a pesticide-mixing facility. (That was not one of our armed services' smartest land-use decisions, even for its time.) He underwent a double mastectomy when he was eighteen years old. It's too late for the handful of men Partain has found who are already dead. And it's too late for Peter Devereaux, otherwise known as Number Seven. Pink is not a color he'd spent much time with. Camouflage, yes.

A native of Peabody, Massachusetts, Devereaux enlisted out of high school in 1980 and was stationed at Camp Lejeune until 1982. He was a field specialist in the Eighth Communications Battalion, and he lived in barracks that used drinking water from the Hadnot Point system. When he came home to Massachusetts after military maneuvers in the Philippines and Hawaii, he started working as a machinist. During the weekends, he pulled in extra income by landscaping, building patios, and moving heavy rocks and dirt around. He was also a serious athlete, running ultramarathons and boxing.

Now he can barely walk. Devereaux was so sick that he couldn't attend the Wilmington meeting, so I called him afterward. He's got a thick Boston accent. "In January 2008, I got breast cancer," he told me on the phone from his home in North Andover. He was forty-five. "My hand had bumped into my chest in the morning. I figured I must have got elbowed playing hoop. Being a guy, you don't even know about breast cancer. I never thought men could get it. But I told my wife, and she made me an appointment to see the doctor." He was diagnosed with stage 3 breast cancer, meaning the cancer had spread to his lymph nodes.

Devereaux was dumbfounded.

"I felt like a freak. I got no breasts, how can this be? I'm a marine, I'm a bad ass, I work out all the time, I ate good, I exercised, stayed fit my whole life, never smoked or did drugs, and you try to figure out how can this have happened?" A month after Devereaux's diagnosis, he received a letter in the mail from the U.S. Marine Corps at the behest of Congress. It stated that he might have been exposed to contaminated water while aboard Camp Lejeune, and it suggested that he and thousands of other marines register at a government website.

"When I got that letter in the mail, within one minute it made 100 percent sense to me that contaminated water was how I got breast cancer," he said. Devereaux found a website started by marines and their families, and soon he was in touch with six other men with breast cancer, including Partain. He agreed to speak out in newspapers and on TV in hopes of reaching more men who may have been exposed at Lejeune. "I gotta let others know, man. I wish they'd let us know twenty years ago, and it could have been a different result for me."

Like many men with breast cancer, Devereaux was diagnosed at a late stage in the disease. His treatment included a mastectomy and the removal of twenty-two lymph nodes, followed by radiation and fourteen months of chemotherapy. "It beat the crap out of me," he said. In 2009, though, he learned the cancer had spread to his spine, ribs, and hips. It was metastatic. "There's no cure this time," he said. Despite being a tough guy, he finds some comfort in the breast cancer community. "You go into all these pink buildings and places for your mammograms and appointments. You're this dude and all these women are looking at you. I meet these women, and they're so much more open and honest and easy to talk to about emotions. Guys, all we talk about are football, eating, farting, and girls. So [these women] really helped. I felt a burden lifted. I wanted to move forward. My goal now is to raise awareness."

But being an expert in combat hasn't hurt. Devereaux wanted to fight the Marine Corps for health-care benefits and help other sickened veterans get them too. Vets can only receive benefits if they have a condition related to their service. He'd been out for over twenty years. Finally, after a two-year argument, he became the second male veteran with breast cancer to convince the government that his cancer was as likely as not linked to the water at Camp Lejeune. To qualify for service-related illness benefits, veterans must prove that an environmental exposure had a 50 percent chance of causing their problems. That may seem like a low bar, but of 3,400 total medical issues brought before the VA so far by former Lejeune residents, only 25 percent have been approved for benefits.

Not all the male breast cancer patients affiliated with Lejeune blame the base for their diagnosis. Take Bill Smith, a seventy-seven-year-old Floridian who was also treated for stage 3 breast cancer. I found him on the website set up by Ensminger and Partain. "I can't say why I got this damned disease," said Smith, who edited the base's newspaper for two years in the late 1950s. "I lived a hard drinking, fun life. I worked the steel mills in Buffalo. I lived at Camp Lejeune. I don't know where it came from. I can't all of a sudden blame the Marine Corps. I don't know and my doctors don't know."

What he does know is that the disease has reformed him from being a self-described swaggering SOB. "I'm not what I was," said Smith, who after his time in the marines worked for many years in advertising on

Madison Avenue. "I was a Mad Man. I was a user of women. I'm not even telling you how many times I was married. I'm not a swinger anymore, not a user. I appreciate women now, and they're so much stronger than men. I went to support groups, I listened to them. I've had the privilege of entering a woman's world."

Most men with breast cancer, especially those who were steeped in a military culture, don't want to talk about it. Partain, though, is as chatty as a schoolgirl in spring. It's why he's such an effective spokesman for his cause. There's nothing girly about his appearance. He describes himself as "a hairy beasty guy," and it's a fair assessment. Not long ago, he convinced Devereaux and a handful of other mastectomized men to pose, topless, in a calendar to benefit breast cancer research.

But underneath his affability runs a deep anger. He is angry that he has breast cancer, angry that the Marine Corps has not done more to apologize to these men or to compensate more of them for their disease. He has vigorously demanded that the Marine Corps turn over more data and notify greater numbers of former residents about the contamination.

As a journalist, I received permission to enter the base and get a tour of its extensive, ongoing $170 million (so far) cleanup mission, which involves everything from oil-eating bacteria to soil-vapor extraction to "pump and treat" stations that oxidize the water's volatile organic compounds into more harmless molecules. Partain, though, said he is not even allowed aboard Lejeune because of standard security protocols. This makes him madder still. So before my base visit, Partain gave me a different tour of his own. We parked across the street, at a dry-cleaners on Highway 24 a few miles outside of Jacksonville. Many commercial enterprises on this strip are named after themes of patriotism or God. The A-1 Dry Cleaners is up the street from Divine Creations Salon and next to Freedom Furniture. Neatly pressed summer camouflage uniforms hang in a row in the window behind a sign that reads, "NAMETAPES MADE AND SEWN ON. 1 DAY SERVICE." This spot used to be called ABC Dry Cleaners, and it was, along with Hadnot Point, a major source of TCE and PCE contamination to the base's water supply.

Partain is a heavy-set man with a goatee and a predilection for aviator sunglasses. He wore shorts and a brown T-shirt that read

"Surf City, USA." The tourist look belied his mood. He pointed across the street to the base, where a chain-link fence and a row of loblolly pines separated the roadway from the base's family-housing neighborhood called Tarawa Terrace. Next to the entrance gate, four bright-yellow pole-stubs surrounded a concrete square the size of a dinner plate. That is now-infamous TT26, a well that supplied Tarawa, where Partain's family lived.

"This is the dry cleaner here, and it slopes toward the river, this way," he said, pointing toward the well. "We are nine hundred feet from TT26. That was the well that was sucking in the plume and feeding the area. They let it pump for thirty years and they poisoned a lot of people. When I look at it and I first saw the monitoring wells, every time I see them I just get angry." Some gulls flew overhead, heading away from the ocean.

"I lived on Hagaru Road until I was four months old. I looked normal and everything appeared normal at first," said Partain, wiping some sweat from his face. "It's every woman's worst nightmare, that something they can do when they're pregnant can affect their unborn child. I've seen it when I talk to the mothers and they learn their child was poisoned and affected. I saw it in my mother's eyes, the most heartbreaking look, despair, that I've ever experienced in my life. To look in my own mother's eyes and see the realization that while she was pregnant, she drank something that harmed her child. I was forty years old when I saw that look. Part of me wants to go on base and show my family, my youngest daughter. She keeps asking me, 'Is what's happening to you going to happen to me, Daddy?'

"I don't want these things burning in my head, but I don't want to stick my head in the sand either. I don't want to forget about it. I have to understand it."

13
ARE YOU DENSE?
THE AGING BREAST

Death in old age is inevitable, but death before old age is not.

— RICHARD DOLL

Most of the time, breast cancer is a disease of grandmothers. At the time my grandmothers got sick, reproductive cancers were not openly discussed. My mother's mother's mastectomy was obvious, though, as a sort of chasm under her matronly dresses, and it loomed large in my childhood imagination. I never knew my father's mother. He was only nine years old when she became ill in Richmond, Virginia. For many years, she would go in and out of the hospital for surgeries or radiation therapy until she died in 1961. To this day, my father loves morning time, because that's when his mother was happiest and strongest, singing in the kitchen and working in the garden. He never gleaned what kind of cancer Florence really had, and it's even possible she didn't know. I'd always heard she died from sort of stomach cancer, and it was only recently that I'd learned it might have been ovarian cancer, which is genetically related to breast cancer. I pursued it with my dad. "The information I got was always filtered through protective layers," he said, fifty years after the fact. "They tried to keep hidden the information that she had cancer. The doctor believed that no patient should ever be told they have cancer." I asked my aunt. "Well, I believe she had some sort of intestinal cancer," she said.

I sent away for my grandmother's death certificate from the Virginia Division of Vital Records, hoping it would have clues. It did. Immediate cause of death (A): malnutrition. Due to (B): metastatic cancer. Due to (C): pseudomucinous carcinoma. I asked my doctor about this diagnosis, and she said, yes, most likely ovarian cancer.

Because of its genetic link to breast cancer, ovarian cancer is also of interest to breast cancer researchers. When breast cancer runs in families, ovarian cancer is often lurking as well. Together, they form a dismal couplet called inherited breast-ovarian cancer syndrome. I'd heard that my great-grandmother, Florence's mother, had also died of cancer, but again, no one was sure what kind. My father had been told it was abdominal cancer, another likely euphemism. Off I wrote to the Will County, Illinois, Clerk's Office, Division of Vital Statistics for her death certificate. I learned that Anne Higinbotham died in 1930 at the age of fifty-eight. "Principal cause of death: Cancer of Lung. Other contributory causes of importance: Cancer of Breast."

Bummer. Two generations in a row of related cancers, only two generations removed from me. Plus a grandmother on my mother's side. After I received the death certificates in the mail, I pretty much ran to see a genetic counselor. I knew the odds if I inherited a mutation in the BRCA genes: up to an 80 percent chance of developing breast cancer and a 45 percent chance of developing ovarian cancer. Shonee Lesh listened to my family history and made a chart full of circles and squares that resembled a geometric child's puzzle.

"My job is to look for patterns," she said. "On your mom's side, there are cancers all over the place, but they don't line up for major concern. It's your grandmother and great-grandmother on your father's side that are the concern. BRCA is the most probable explanation. It's high enough for us to have this conversation and for you to be tested."

My insurance company, however, disagreed. It would pay for testing if I had a first-degree relative with breast cancer (mother, sister), but not for grandparents, even two in a row. The BRCA genes are patented by one company in the United States, Myriad Genetics, and it has decided the test to decode the BRCA1 and BRCA2 genes will cost its customers about three thousand dollars. It's expensive enough to make insurance companies fairly ruthless about it.

BRCA genes are most commonly found in eastern European Jews, with about one in forty carrying the most common genetic errors (in the general population, the rate is about one person in five hundred). But my Higinbotham foremothers were not Jews. They might have carried other mutations in the BRCA genes, such as one known to have arisen in Iceland in the mid-sixteenth century. This mutation, called 999del5 BRCA2, developed because of a missing piece of DNA in a single individual who enjoyed some reproductive success. Or my grandmothers could have inherited one of the seven hundred other distinct cancer-causing variants found in the BRCA genes among Dutch, Germanic, French, Italian, British, Pakistani, or French-Canadian populations.

The BRCA genes are the most common and deadly of the genetic variants, but there are numerous others — some discovered, some not yet — my grandmothers could have inherited. In families with histories of breast and ovarian cancer, less than half of them have BRCA mutations. It's also remotely possible that my foremothers, coincidentally, developed their own, unrelated, non-inherited mutations. In total, only about 10 percent of breast cancers are believed to stem from a heritable gene flaw.

The average lifetime risk of breast cancer in the United States is one out of eight, or 12.2 percent of women who reach the age of ninety. When Lesh plugged my risk factors into something called the Tyrer-Cuzick Risk Assessment Model, it calculated my risk at 19.8 percent. Lesh told me that when an individual's risk reaches 20 percent, doctors recommend aggressive screening, such as annual or semiannual pelvic ultrasounds (for detecting ovarian cancer) and semiannual breast MRIs in addition to mammography. But absent BRCA testing, the results of which could push my magic risk number way up, I, like most women, would be more or less on my own.

If cancer is a disease of aging, the older we get, the more vigilant we need to be. It seemed like a good idea to understand the risk factors. Age and family history may be the major ones, but as I was learning, they're not alone. Other standard risk factors are early puberty, late menopause, obesity, older maternal age, a record of a previous breast abnormality, and race (white women have a slightly higher risk than African Americans and a considerably higher risk than Asians or Hispanics). But — and here's the disconcerting part — most people who get breast cancer have few of these risk factors, other than the big buckets of age and race. A stunning majority of women with breast cancer — 90 percent — have no known family history. Equally perplexing, most women with the risk factors, even a bunch of them, still never get breast cancer. In other words, the standard risk factors are fairly useless. We still don't really know what causes breast cancer.

Obesity is a good example of how confusing things can get; it's a risk factor for postmenopausal women, but oddly, a protective factor for younger ones. Other risk factors have been or are being considered for inclusion in risk models, and if you've read this far, you know some of them: radiation and chemical exposures, alcohol consumption, a high-fat diet, use of birth control pills, hormone replacement therapy, and nationality. Women in the United States and the Netherlands have the highest rates in the world. Japan has among the lowest. Scotland is middling. Interestingly, women in China get the disease, on average, ten years earlier than their counterparts in North America. Lately, a newish risk factor has emerged, and it's not one often thought about: breast density. If you haven't been clued in, you're not alone. One fifty-year-old friend told me she'd recently returned from a mammogram. The radiologist told her she had very dense breasts.

"Thank you!" she burbled, thinking it a compliment. I had to explain to her that the doctor was not commenting on her firmness, which is, it must be said, admirable. I told her that density is a measure of the ratio of fat to glandular tissue. She looked decidedly dispirited. Not only that, I continued, but dense breasts make reading mammograms difficult, and women with dense breasts are at higher risk for breast cancer, a double whammy. Now she was glaring at me. I changed the topic. But I'll say more here. Two-thirds of women go into menopause with dense breast tissue, and one-fourth retain it afterward. Women with the densest breasts are believed to have a four- to five-fold greater cancer risk than their peers, making density the biggest risk factor for cancer after age. It's also the biggest risk factor you've never heard of: 90 percent of women do not know if they fall into this category.

In a better attempt to know my breasts and foretell their destiny, I hied my aging, American self down to Dallas. There, I met Dr. Ralph Wynn. He is the kind of man you'd want to be your radiologist, should you ever need one. A soft-spoken Texan, he's kind, careful, and very experienced. He's been reading murky mammograms for over twenty years, and has put in time at some of the best cancer centers in the world. He can find the proverbial needle in the haystack, seeing minute "disruptions" in impossibly hazy fields of white-and-gray X-rays and sound waves. Although Wynn has recently been named director of breast imaging at Columbia University Medical Center, I was lucky to catch him while he was still practicing at the University of Texas

Southwestern Medical Center and overseeing the country's only commercially available 3-D breast ultrasound machine. I didn't want to miss out.

It's not easy to see inside breasts. If it were, mammograms wouldn't miss 20 percent of all tumors. MRIs are better, but they require getting injected with a dye and spending thousands of dollars, not to mention enduring forty-five minutes of lying immobile in a tube the size of a small sewer main. They are the domain of high-risk women. Three-dimensional ultrasounds could be a decent compromise. They have been used in Sweden for years, but they are new to the United States and insurance will not pay for them yet. Wynn is participating in a national study to compare their effectiveness to mammography in a bid to get them more widely used. The general consensus is that ultrasound picks up more tumors than mammography, but it also picks up more lumps and shadows that are not cancers, the so-called false-positives that are the bane of women, insurance companies, and government task forces. Wynn wanted to know how many lives could be saved by this technology and at what cost. My interest, though, was how ultrasound can draw back the curtain on how my breasts are changing as they enter middle age.

For his study, Wynn enrolled several hundred women with dense breast tissue. He said he'd be happy to examine me as a test case. He asked me to send him two sets of mammograms — my oldest and most recent films — before arriving.

On the appointed spring day, Wynn met me in the lobby of the modern Seay Building on the sprawling University of Texas campus north of downtown. We walked past the immense lobby sculpture of shiny orange globules stretching to the high ceiling. "I think it looks like sperm," said Wynn. I was thinking the same thing. He introduced me to Robin Eastland, the technician who would operate the 3-D machine, called the Somo.v. A cheerful Texan in her mid-thirties, Eastland led me to an exam room on the third floor and handed me a gown, which was, naturally, pink. When I was settled in, she told me that unlike mammography, this machine would not squeeze my breasts in a vise grip. The worst part of the procedure would be the cold gel.

I lay down on an exam table, and Eastland parted aside my gown. She held a tube of sonography gel over my right breast.

"Ready?" she asked.

I nodded, and the tube belched out a cold

substance resembling Elmer's glue. Eastland smeared it around. Then she maneuvered a book-sized square attached to a mechanical arm above my breast. The bottom of the square held a disposable chiffon-like screen that compressed and conformed to the outer half of my breast. She pressed a button, and an automatic transducer on the other side of the screen moved down my breast like the rollers on a massage chair. The machine sent high-frequency sound waves into my tissue, recording the time it took them to bounce back. (Sonography is also used to help boats find deep-sea fish and to measure fetuses in the womb.) In my breasts, when the waves encountered a change in tissue density, such as from a cyst or rib, the signal hesitated, and the object's size and location got marked as a dark color on a computerized 3-D map. The whole breast map took several minutes.

When we finished, we found Wynn in his reading station, which reminded me of the scene in *The Matrix* where Keanu Reeves meets the sentient machines. Six large screens surrounded a couple of office chairs in a small dark room, each flashing pictures of images from mammograms and ultrasounds. "I spend most of my time alone, sitting in the dark," explained Wynn, who has close-cropped hair and round wire glasses. Now that he said it, I saw that he was a little pale. I wanted him to drink an Arnold Palmer and go play golf in the sun like normal doctors, but first I wanted to see my pictures.

One neat thing about digital 3-D ultrasound is that the CAD software can produce both coronal slices — like individual cuts of deli ham — or the whole ham hock from any angle you want. If you go for the slices, each one offers a view about two millimeters thick. Since the average tumor is twice that size, it will probably show up. The images aren't as crisp as a mammogram, but the contrast is better. On a mammogram, breast tissue looks like a big uneven clump of snow, while a tumor might look like a cotton ball. On an ultrasound image, a tumor looks like a very dark patch on a bed of somewhat lighter patchiness. Wynn's job description, then, falls somewhere between reading tea leaves and looking for eagles flying across a night sky.

First Wynn pulled up the rotating 3-D image of my breast, slightly flattened by the rollers and appearing taller than it was wide.

"It looks like a fat piece of French toast," I said.

"Or croque monsieur," he countered.

"Panini."

"Grilled cheese."

We'd now established we were hungry. To get through all the slices for both of my breasts meant reading about five hundred images. Wynn flew through these with his mouse wheel, moving from the nipple to the chest wall. He could have been Captain Kirk piloting at high speed through a remote, hazard-filled galaxy. "Your breast tissue looks nice and homogeneous," he said. I'm relieved. But then he slowed down through some honeycomb patterns and told me my tissue is fibrocystic, especially toward the outer edges. This can be a risk factor for disease. He kept scrolling. "Here we can see ductal structures radiating from the nipple." They looked like fuzzy spider veins of varying thickness, indicating that some contained cellular fluid.

Not all of my milk ducts were visible. Some, explained Wynn, had regressed with disuse. That notion made me feel like an expired dairy product. It had only been five years since I last lactated, but already my glandular cells were being replaced by fat cells. (If I get pregnant now, in my early forties, this process would quickly reverse.) This surprised me; I knew breasts grew less dense after menopause, but I didn't realize this process would be so evident so soon. Wynn pointed to the light parts of screen. "All of this here and here and here and here and here and here is fat," he said.

To better show me the changes in my breasts over time, he uploaded on the opposite computers the CD my doctor had sent of my previous mammograms. First he pulled up an image from my last mammogram, taken six months earlier. Here the colors were reversed; glands are white, fat is dark. "The more white, the more dense," he explained, pointing to a portion of the image. "The fat content in here makes up at least 75 percent of this whole area. It may look denser to the untrained eye because of superimposed parenchyma." (A quick refresher from chapter 3: the breast is made of three major things: fat, gland, and stroma. The gland is sometimes called ductal, parenchymal, or epithelial tissue. The stroma is the extracellular universe surrounding the gland and supporting it. It includes collagen, growth factors, and proteins and also looks whitish on a mammogram.)

Because of my family history, I got my first baseline mammogram when I was just thirty-three, and got another after I turned forty and was finished with pregnancy and breast-feeding. The point of a baseline is to have something for radiologists to compare with later images. Wynn pulled up my first mammogram. It's from an older machine, and fuzzier. "You can see there was fat then but there was a lot more white," he said, pointing to the screen. "And so you're progressively depositing more fat as the fibroglandular and ductal system atrophies. It's a good thing you don't have denser breasts now. For your age, you have tissue that's appropriately regressing and you have progressively more fatty breast tissue." In the early film, my ratio of fat to gland was about 40-60 whereas the ratio ten years later was the reverse. The verdict: moderately dense tissue, not high risk.

I don't love it when people start sentences with "for your age." But with breasts, the march of time is inexorable. Most women's breasts are like mine, losing dense tissue as their reproductive years wane. Breasts are considered "very dense" if the gland and stroma remain, taking up 75 percent or more of the breast. No one is sure why some breasts are denser than others, but it tends to be hereditary. There's a big search on to identify these genes in the hopes of someday linking them to cancer and targeting them with drugs.

Women tend to have denser breasts if they've never had children. Hormones also influence density. Menopausal women taking hormone replacement therapy develop denser breasts almost immediately. If they take tamoxifen (an anti-estrogen drug used in cancer treatment), the mammary gland retreats and gets replaced by fat. Not everyone should start popping tamoxifen, but it proves that drugs can change your breasts, and fast. Some studies show that wine drinkers have denser breasts, as do smokers and women eating high-fat diets. These things may turn on or off genes in ways that promote inflammation, growth, or instability in glandular cells.

In this way, density stands in for breast cancer risk overall. If a woman is postmenopausal, she can reduce her risk by eating well and exercising and by not drinking excessive alcohol or smoking. Unfortunately, though, these gains are small. It appears that by the time a woman reaches menopause, her cancer destiny is mostly laid out by some mysterious combination of her genes, the pattern of growth taken by her breasts, and the accumulated damage (or lack thereof) to her cells over many decades. Menopause is simply the end zone in a long game of chicken between breasts and carcinogens. By this stage of life, it's too late for a woman to change the things that may have set her down a particular path: the childhood exposures, her reproductive history, the hardiness of her genes. New exposures, such as to hormone therapy, may put her over the edge. But her cells will keep aging no matter what she does, and as they do, they'll collect more mutations. Most women take hormones without a hitch. The risk — a doubling in deaths from breast cancer — sounds bad, but it is the equivalent of about two additional breast cancers per year for every ten thousand women taking hormones. It's enough of an effect, though, that when a third of hormone users quit following the study results of 2003, the U.S. breast cancer rate noticeably declined.

I was increasingly learning that the whole blame-your-lifestyle approach to understanding breast cancer is problematic. In a way, it presents an excuse not to probe into the deeper reasons for disease. As environmental historian Nancy Langston put it, "Traditional medicine and public health practices have been reductive, focused on individual risk factors for disease." Instead, she argues, we need a more ecological understanding that explores how genes and the environment interact to compromise our immune system in the first place. Ultimately, we should be asking and answering, *Why* do some women have dense breasts? Is there anything we can do to prevent or lessen the impact once it kicks in?

In lieu of that understanding safeguarding our breasts any time soon, I figure knowing our tissue density can at least help us make more informed decisions about the choices we have left in middle age, but with the knowledge that those choices are imperfect. Women with very dense breasts might want to avoid taking additional hormones if the benefits aren't worth it for them. They might want to lobby their insurance companies so they can get screened more often, using a greater variety of technologies like the 3-D machine to boost their odds of catching problems. Mammograms might work pretty well for most older women, who tend to have low-density breasts and slow-growing tumors. Even in this group, however, the benefits of early detection are debatable, because it's likely that many of these tumors would not be lethal. The statistics get more depressing on the effectiveness of mammograms for women between the ages of forty and fifty, who often have more aggressive, fast-growing tumors that are harder to spot. Recall the enormous flare-up in 2009, when the U.S. Preventive Services Task Force reviewed the data and recommended against the decades-old policy of women under fifty getting routine mammograms (later, the panel backpedaled to say screening decisions should be made by patients and their doctors).

Here's the sorry and under-sung fact: mammograms for my age group are lousy. Thanks to time spent in Wynn's flight deck, I now know why: we still have too much white stuff (the dense glands) that X-rays can't see through. A 20 percent failure rate is just not good enough. But what especially rankled the task force were the costly false-positives. Better to do none at all, it implied, until your breasts fatten up. That recommendation wasn't the only blow the task force dealt. There was another that drew far less attention: women should no longer be taught how to perform breast self-examination, known as BSE. Equally disheartening, the task force went on to say there wasn't even enough evidence to recommend clinical breast examination, the kind performed by your doctor during an annual checkup.

Like many women, I wasn't liking the options left by the task force. We've all heard that early detection can be the key to surviving this disease, at least for some if not all tumors. But how, in women under fifty, is a growing tumor supposed to be detected without mammography or people looking for it? Where I live in Colorado, fully one-third of all breast cancers occur in women under fifty. Put together, these two recommendations meant that we might as well just take up voodoo and buy a Magic 8 ball.

As blogger Leigh Hurst put it, "Wow — are you kidding me? How can this be? A BSE is what saved my life." Hurst found her tumor when she was thirty-three and has gone on to promote breast self-awareness through a hip website called Feel Your Boobies. It's a well-recognized fact that most breast cancers are found by women themselves, not by mammograms. Often, this happens by accident, not during a formal search-and-destroy mission.

I learned that the task force's BSE guidelines were based on two large studies, one in China and one in Russia. Those studies compared women who were taught how to do BSEs and did them, with women who did nothing, and found discouragingly similar death rates from breast cancer. At the same time, the women who performed BSEs found more false lumps.

But a number of other experts have criticized those two studies as flawed, saying, for example, that the women in China received inadequate training and that the Russian study ran out of money for follow-up. Other studies modestly support BSE, including one in Canada, which did find a lower death rate in women who were well trained. A recent study from Duke University found that mammograms, MRIs, and BSEs were equally effective in finding tumors in high-risk women. For women at highest risk of breast cancer, mammography may actually be hazardous, since faulty BRCA genes make breast cells more sensitive to the damage caused by radiation used in the procedure. For these women, BSE might actually be their best option.

The strongest argument against BSE is that it's difficult to do properly and requires training. For a large population, it's simply unrealistic, according to Dr. Russell Harris, who served on the U.S. task force and supports the recommendation. But for a motivated individual, BSE could be your best friend. As Lee Wilke, a breast surgeon and the author of the Duke study, told me, "BSE turns out to be only as good as the person doing it."

I suddenly wanted to be very good at it. There would be no more half-assed shower gropings. I was going to learn to do it properly. I bought a fake, silicone boob (forty-eight dollars from Amazon.com) and considered where best to stick it on my chest. It wasn't just any fake boob; this one came with lumps and bumps designed to mimic the nodularity of real breasts. It also came embedded with a number of "tumors," or harder bits of plastic of various sizes and at various depths. It felt almost disturbingly real, with a squishy nipple and smooth skin. Manufactured by a company called MammaCare, this model was designed to teach women how to perform "tactually accurate" BSEs.

Per the instructions, I lay down and placed the cool falsie below my collarbone, which made me feel like a multi-teated mammal. I popped the accompanying DVD into my laptop, which I perched on my stomach, and prepared to enter the world of low-tech, last-resort cancer detection.

MammaCare is considered the Harvard of BSE trainers. Its squishy silicone booblets are used in the Mayo Clinic and in medical schools throughout the country. Company cofounder Dr. Mark Goldstein told me they were designed in a university lab after almost laughably painstaking research into "pressure load curves of the human breast." Goldstein is considered a sensory scientist, a man who believes that we can train and use our senses to work like finely calibrated machines. He told me his father ran a metal fabrication company and could judge the correct width of sheet metal within a hundredth of an inch, using his fingertips. One night this man happened to detect a three-millimeter tumor in the breast of his wife (Goldstein's mother). Most tumors are ten times that size by the time a woman or her partner finds them. Goldstein wanted to create a training program that was simple, thorough, and effective, and that ordinary women could use. He said BSE, performed right, is as accurate as mammography, especially in women under fifty.

"We can take someone who can't find a marble on a table and teach them to detect a three-millimeter tumor inside a breast," he said.

I was ready. I pressed play. A no-nonsense woman with an early-1990s hairstyle introduced the concept of the "vertical strip" search pattern. Goldstein calls this "mowing the lawn." The circles of yesteryear are clearly no longer in favor. Following along with the video, I proceeded to feel up my model along these lines, using the fat pads on my three middle fingers to create a small dime-sized zone. I dutifully applied three pressure depths — surface, medium, and deep — at each spot on the grid. I immediately detected two small, hard "tumors" on the left side and one on the right. During the review a few minutes later, though, I found out that I missed two others, including a big one deep under the nipple. To feel those, I had to press down much more firmly. If my model had been a water balloon, I'd have popped it. I was unsure whether I'd have the guts to press that hard on the real deal, and I was right.

When it was time to trade the model for my own breasts, I could tell right away that things are much more complicated in flesh and blood. If the model represented the geography of rolling tundra, my body felt more like the great Himalayan upthrust, complete with granite, lakes, ice, snow, and the occasional civil war. It was harder to tell what was going on or where a cancer might lurk. And it was harder (and painful) to push down very far through all my natural ropy tissue. If I were to develop cancer, I'd have to hope for shallow tumors. Also, I have to admit, it's frightening as hell. What *are* all these bumps? And the exam takes some time, about seven glacial minutes per breast when you're starting out. Discouraged, I called Goldstein to ask for tips. He told me that the more I practice BSE, the better I'll get at telling what's normal and what isn't, especially if I do the exam at the same time every month, ideally at the beginning of my cycle before the late-month progesterone-hit makes things even knottier in there. "The fingers remember," he reassured me. "They operate brilliantly, but they need to be used. You can't sit down at the piano and start playing Mozart." He also reminded me that having my breasts squished between mammography glass hurts even more. Good point.

I believe Goldstein; I believe that it's possible and important to learn to do this well. I would like to think I will do BSEs, if not every month, then at least a few times between mammograms. But I also have to acknowledge I was never great at practicing piano, and I recognize I might not be destined for BSE virtuosity. I called William Goodson, a San Francisco–based breast cancer surgeon and researcher and another proponent of BSE. He told me that just getting to know one's own breast geography is a major accomplishment. For women who can't bring themselves to conduct the full-on regular BSE, just better breast awareness is a big step. No one knows your breasts like you do. "It's useful to have a woman become familiar with her breasts, to be aware of any changes. You've got to sit down and look at them. And don't only look for lumps. Many cancers feel like a more irregular area, where the skin doesn't move right or feel quite right."

This much, I hope, I can handle. If I've learned anything, it's that cancer detection is as much art as science. BSE isn't perfect, and it's not going to work for everyone. But my foray into self-monitoring has convinced me that I can undertake some useful reconnaissance. As cancer survivor Hurst put it, "We're lucky our breasts are on the outside of our bodies where it's possible to become familiar with how they feel. We're not talking about our lungs here."

My best advice to you, dear reader: know thy breasts.

14
THE FUTURE OF BREASTS

> The world is too much with us; late and soon, Getting and spending, we lay waste our powers
>
> — WILLIAM WORDSWORTH

It was mid-August, time to get out of the labs, turn off the computer, and head for the mountains. I had a good excuse: a fund-raiser for breast cancer research. I guilted my friends into donating some money, and I promised I would try to haul myself up three fourteen-thousand-foot peaks in one day. Thankfully, they are very close together. My team included five women and one very tall man wearing an Indiana Jones hat. He would be our beacon.

We were all climbing for different reasons. Lisa and Steve were climbing in honor of her mother. Natasha and Cindy were climbing for the camaraderie and a good cause. I was climbing in memory of my grandmothers, and for my daughter, in the hope that we may learn how to prevent breast cancer in time for her. Then there was Sherrie, whose daughter, Lesley, hadn't been so lucky. She'd recently been diagnosed with breast cancer at the age of thirty-four.

Lesley, round faced and red cheeked, bundled in a puffy down jacket and head-scarf, walked us to the trailhead at dawn. She distributed our T-shirts, which read, "Save Second Base," then wished us well, and gave her mom a tight hug.

Sherrie shouldered her sizeable pack and thundered up the trail. At sixty-seven, she is built like a greyhound, sleek and muscle ripped. When she's not climbing mountains, she's a triathlete and ski instructor.

"What are you listening to?" I bellowed into her headphone pods, expecting a flavor of classical.

"Red Jumpsuit Apparatus," she bellowed back. "I like alternative."

A couple of hours later, at the summit of Mount Democrat in Colorado's Mosquito Range, I learned that Sherrie's nylon pack was like Mary Poppins's carpet bag. Out came a smorgasbord of honey energy gels, goo packs, and other electrolyte snacks, enough for everyone for this and two more peaks.

Newly fortified, we admired the view. It was a glorious Rocky Mountain day: bluebird sky, granite peaks, patches of white snow sparkling in the sun, triangular green trees far below. For 360 degrees, from 14,148 feet, it was hard to discern any houses or roads. For a moment, it was easy to feel the world was as nature made it.

But then, to the West, what at first looked like a lake revealed itself to be a tailings pond. It was too rectangular and a strange color of turquoise. This was part of the Climax mine, owned by multinational giant Freeport-McMoRan Copper & Gold. Until recently, the mine was the world's largest source of molybdenum, a trace element used to harden steel. Now it's part of a notorious Superfund site. It was hard not to think about the cost of this pollution to the mountains and to our national budget, and then, directly or indirectly, to our breasts.

I'd rather just look at luminous, neighboring Mount Lincoln, no tailings in view. But this is the cropped way we've seen breasts, and it hasn't done them much good. Scientifically, medically, and culturally, we have preferred to think of them as disembodied objects, removed from the rest of the human body and separate from the rest of nature.

In the early days of cancer treatment, surgeons believed if they could just hack off the tumor-ridden breast, the patient would survive. Alas, even radical mastectomies didn't usually solve the problem. By the time the cancer was found, it was already invisibly wired to distant parts of the body. But the failure to recognize this cost many women needlessly painful and debilitating procedures. We were also slow to understand that pollutants could end up in breast cells and in breast milk. By insisting that breasts are sexually evolved and relegated to a sexual destiny, we have encouraged women not to value breast-feeding, and sadly, often not to value their normal, natural bodies.

Breasts have only slowly offered up their secrets, and we have been too distracted by their beauty to look very hard. As we've learned, breasts aren't static, or pneumatic. They are always changing.

New bras, new lumps, new pride or despair, new glory, new fear.

Women often talk to me about their breasts. They'll tell me that they're donating some extra breast milk to a friend in need, or that their brother had breast cancer, or that their breasts are uneven. I'll tell them about how humans are so unique to have rounded breasts all of our adult lives, how chimps' breasts go back to being flat as Frisbees after nursing. "That happened to me!" they might joke. We'll move on to other topics, like tonsils or tornadoes or *Mrs. Dalloway.* Sooner or later, though, the conversation usually comes back around. Friends and acquaintances ask where I stand on annual mammograms. Am I worried about false-positives? What about the radiation? I tell them the truth, which is that mammography is an archaic and deeply flawed technology, barely improved over fifty years. There are many false-negatives and false-positives, and, yes, there's radiation, the only proven environmental cause of breast cancer. We can invent the Internet but not something better than blasting ions into our boobs using forty pounds of pressure?

So what do breasts need? ask my friends. Is there hope? Breasts desperately need a rosier future. They need a safer world more attuned to their vulnerabilities, and they need good listeners, not just good oglers. Breasts have some terrific allies, such as mountaineer/chemist Arlene Blum, who is fighting to get flame-retardants out of baby products and out of breast milk, and Susan

Love, the brash surgeon who started a nonprofit to fund experimental research using a million human volunteers. Among the promising things Love is looking at: cheap tests of breast fluids to identify women at high risk of cancer. Then there is Rachel Ostroff, research director at SomaLogic. The Boulder-based biotech start-up is trying to identify proteins and enzymes in a routine blood sample that indicate the presence of an early breast tumor. Ostroff, whose sister died of breast cancer, calls these tiny protein biomarkers "the voices of life." "We want to find the early evidence of disease when a cure is possible and simple," she said.

These are heartening advances. The medical community is getting better at treating women with breast cancer. In 1944, only 25 percent of women with breast cancer survived ten years; by 2004, the number had increased to 77 percent. Ultimately, though, diagnostics and treatments are already measures of defeat: the tumor has arrived. We can save many women from dying of breast cancer, but that's not necessarily going to save their breasts. To save breasts — and to spare women the particular agonies of this disease — we need to think more about the bigger picture of health and, ultimately, prevention. Yet surprisingly few national research dollars — about 7 percent of the budget of the National Cancer Institute — are spent on prevention, even broadly defined to include early screening. The cynic in me would point out that actually it's not that surprising; there is little money to be made in preventing breast cancer compared to screening for it or treating it.

"It's always been my belief that prevention will help us so much more than a cure," said Saraswati Sukumar, a Johns Hopkins biochemist. With some help from the Dr. Susan Love Research Foundation, she's pioneering "chemical mastectomies," injecting chemotherapy through the milk ducts. Currently, the idea is to kill early, proliferating cells and allow women to keep their breasts. But eventually, says Sukumar, the procedure might become part of regular maintenance in high-risk women, akin to regularly cleaning out the pipes, Drano-style.

Call me a dreamer, but I'd rather have healthy pipes to begin with. This is where prevention gets trickier. On a personal level, it means making vexing choices: drinking less alcohol, seeking alternative menopause treatments, exercising (a lot), scrutinizing labels, and reducing exposures to toxins and endocrine disruptors. These measures will take us only so far. A better and more successful approach would be a societal one, in which industries have incentives to design safer products and make healthier foods, and governments adopt a commonsense and rigorous approach to testing and regulating chemicals. Social policies could encourage more breast-feeding and less obesity.

Pharmaceutical companies are already working to dethrone standard hormone replacement therapies. Women may soon be able to take a pill that allows estrogen to reach bones and skin but blocks breast cells from picking it up. Similar drugs are in trials now. I'll admit, though, that I'm skeptical of these solutions. They remind me of altering an ecosystem to benefit one species, such as rainbow trout or Roundup-ready soybeans, only to see harms pop up unexpectedly downstream or downwind.

When it comes to breasts, the ecosystem metaphor is apt. The new sciences of environmental disease and epigenetics are redefining the very notion of human nature. They're challenging us to recall an ancient belief system that says we are deeply con-

432

nected to our environment. Twentieth-century medicine had us believe our DNA was our destiny. Now we understand our DNA was built to bend. The pendulum of science is swinging away from the preeminence of the genetic code to the surprising power of our soil, air, water, and food. In this current cultural moment that worships technology and throwaway convenience, it's a good time to remember our physical interdependence with the larger world. If breasts are to be saved, their salvation will lie in this recognition.

Much about our environment is better than ever. We have fewer parasites and infectious disease; most of us are protected from extreme weather and food shortages. On the whole, people in developed countries are smoking less and living longer than ever before. But when girls reach puberty earlier, their young lives face new and difficult challenges. Toxins in breast milk run the risk of affecting the cognitive, behavioral, and physical health of our children; and breast cancer will, on average, shave thirteen years off a woman's life. We now understand health to be more than a measure of longevity. Our goal should be to live the best lives we can.

Decades ago, microbiologist-turned-humanist René Dubos argued for an ecological view of health. Health, he said, was not simply the absence of disease but the ability of the body to adapt to dynamic circumstances. Modern life appears to have compromised that ability in many ways. To safeguard our breasts, we need to protect our bodies' biological processes. In this sense, the word *ecosystem* is no longer just a metaphor. Breasts *are* an ecosystem, governed by long-evolved functions, migrating molecules, and interconnected parts. Like every ecosystem, this one is highly adaptable, to a point.

Breasts are our sentinel organ. They offer us a window into our rapidly transforming world and the excuse to steward it better.

Much like the Mosquito Mountains, or free-flowing rivers or polar ice caps, the human breast is a complex, unique, thrilling, beautiful thing, connected to the world in ways grand and infinitesimal. It's an evolutionary miracle that we are only beginning to understand.

The fact that we can now talk so freely about breasts means our blind spots are getting smaller. More women are probing the role that industrial chemicals play in polluting our bodies and our breast milk. Every year, more scientific organizations issue proclamations on the need to research and regulate endocrine disruptors. Last year, an esteemed group of international researchers called on regulators in the United States and Europe to stop ignoring mammary glands when performing studies on lab animals. Those glands, they reminded us, may well be our most informative tissue.

As everybody knows, breasts can speak loudly. When we became aware of chemicals in breast milk, a powerful new lobby of mothers helped sweep DDT and PCBs off the marketplace. The same fate will likely befall brominated flame-retardants. But for scores of other chemicals, the science is young and the regulators often stripped of meaningful power. For the sake of breasts, let's hope both of these conditions change. In the meantime, consumers have a few more options. Soon it will be possible to buy some kitchen plastics that don't stimulate the estrogen receptor. Huzzah! It's a start.

Breast cancer rates declined a few percentage points after 2003, probably because fewer women were taking hormonal replacement therapy. Unfortunately, the decline ended in 2007. Perhaps the HRT effect has gone as far as it will go, or perhaps other factors have started to fill in the vacuum. The search for all the interwoven causes of breast cancer will take a long, long time.

After my grandmother got her radical mastectomy in 1973, a relative asked her how the wounds were healing. She replied she didn't know because she refused to look.

There I was, thirty-eight years later, atop a mountain, considering the blue-green splotch of the Climax tailings pond and popping a cherry gel. Sherrie, my indefatigable hiking companion, pulled out her water bottle and camera, and then, slowly, she unfurled her homemade yellow sign for the summit photo: "For Lesley — and for Dan and Alli and for Mom, Bernie, Debbie, Maggie, Dad, George, Sheila, Stef, Staci, Shawn, For Courage . . . For Hope, and For Love."

I'm with her. I didn't have a sign, but I could have easily shouted down the soft flanks of the Mosquito Mountains that I was there for my daughter, my friends, and my grandmother who refused to look. It's for them that I wanted to.

ACKNOWLEDGMENTS

It can be awkward explaining to people that you're writing a book about breasts, but this project was met with great enthusiasm by many talented and helpful people. My agent, Molly Friedrich, is, like breasts, a force of nature. Her immediate and unwavering faith, along with that of Lucy Carson, propelled me and kept me afloat. I was lucky that Norton's Jill Bialosky believed in this project from the beginning. This book was much improved by her editing and insight. I'd also like to thank the hard work of the copyeditor, Mary Babcock, and my publicist, Erin Lovett.

As I wrote this book, I kept pinching myself that I could be so lucky to dive into such a fascinating subject and the rich, varied research surrounding it. It was a privilege to engage with so many brilliant, dogged minds. I can't list them all here, but I'm full of gratitude to the many people who took uncommon time to spend with a reporter. Numerous scientists and doctors welcomed me into their labs, offices, and conferences and sometimes read portions of the manuscript for accuracy: Olav Oftedal, Malcolm Pike, Irma and Jose Russo, Frank Biro, Larry Kushi, Zena Werb, Dan Sellen, Alan Dixson, Barnaby Dixson, Pepper Schedin, Patricia Hunt, Shelley McGuire, Peter Hartmann, David Newberg, Bruce German, Patricia Adair Gowaty, Ralph Wynn, Susan Love, Dixie Mills, Bernard Patten, Tom Biggs, and Michael Ciaravino and his patients. A handful of scientists helped me formulate interesting and telling ways to test my breasts, body, and home environment for various chemical substances; I'm grateful to Ruthann Rudel, Julia Brody, Heather Stapleton, Arlene Blum, Sonya Lunder, Andrea Kirk, Åke Bergman, Olaf Paepke, Arnold Schecter, and, at Axys Analytical Services, Barbara Carr. For their generous time and candor with their personal stories, a special thanks to Michael Partain, Pete Devereaux, and Timmie Jean Lindsey. The experts frequently set me straight, but any remaining mistakes are entirely my own.

I'm fortunate to have many supportive colleagues and friends who offered wisdom, edits, pep talks, babysitting, and the occasional enlivening Thai taco over the years of this project. Lisa Jones and Hannah Nordhaus were not only elegant role models but also went the extra mile with their red pens. Thanks to Ginny Jordan for inspiration and support, and thanks also to Hanna Rubin, Hillary Rosner, Melanie Warner, Claire Dederer, Tracy Ross, John Heyneman, Sandra Dal Poggetto, Brian Kahn, Caroline Patterson, Michelle Nijhuis, Paolo Bacigalupi, Page Pulver, Carin Chow, Susan Moran, Rachel and Jeff Walker, Deborah Fryer, Bonnie Sue Hitchcock, Curt Pesman, Andrea Banks, Auden Schendler, Anders Halverson, Peter Heller, Dan Baum, Beth Judy, Rebecca Stanfel, Rick Newby, Edward Lewine, Todd Neff, Joe Sorrentino, Sean Markey, Jim Levine, Betsy Tabor, Laura Tabor, Noah Harwood, Metta Gilbert, Lauren Seaton, Barbara McGill, and Danielle Garson. My friends and relatives, despite suffering bouts of neglect, offered sage counsel and various assistance. Thanks especially to Pamela Geismar and Pete Friedrich, Mara Rabin, Ann Vileisis, Margaret Nomentana, John and Galina Williams, Jamie and Wendy Friar, Terry and Joe Williams, Peter Williams, and Herr Professor Joe Williams Jr., whose scientific genius gives me inspiration and who delivered a much-needed private tutorial in statistics. Penny Williams took the Grandmother Hypothesis to new heights, moving in for weeks at critical times to feed and otherwise nurture us. Thanks also go to the occasional research assistance provided by Breanna Drexler, Jordan Wirfs-Brock, and Keirstin Kuhlman.

I was fortunate to have financial and administrative support for this project from the Ted Scripps Fellowship in Environmental Journalism and the Center for Environmental Journalism. At the University of Colorado, special thanks to Doña Olivier, Tom Yulsman, and Len Ackland. I'm grateful to the Lukas Prize Project Awards committee at Columbia University for honoring this manuscript as a distinguished finalist.

A number of relatives, friends, and friends of friends took me in during reporting trips or offered studio space so I could write: Betsy and Andy Baur, Ann Skartvedt and Mark Burget, Chuck and Molly Slaughter, Terry Hasshold, Philip Higgs, Chris Todd, Michael Kodas and Carolyn Moreau (and Otto, of course), Julie Frieder and Charlie Stanzione, Beth Conover and Ken Snyder, Garrett Finney and Sarah Newbery, Cyane Gresham and Alan Bayersdorfer, Violet

Wallach, Jon Hoeber and Jenn Leitzes. I hope to return the favors.

Many fine magazine editors and teachers over the years have taught me much and indulged my curiosity, mammary and otherwise, in ways that supported this project: Ed and Betsy Marston, William Cronon, Fred Strebeigh, John Wargo, William Kittredge, Elizabeth Hightower, Emily Bazelon, Jennifer Rainey Marquez, Megan Liberman, Vera Titunik, Jamie Ryerson, Jonathan Thompson, Amy Linn, Toni Hope, Peter Flax, George Black, Laura Wright Treadway, and Alan Burdick.

Several books stand out as having particularly inspired, influenced, and cowed me: Natalie Angier's *Woman,* Marilyn Yalom's *A History of the Breast,* James S. Olson's *Bathsheba's Breast,* Siddhartha Mukherjee's *The Emperor of All Maladies,* Sandra Steingraber's *Having Faith* and *Living Downstream,* Rick Smith and Bruce Lourie's *Slow Death by Rubber Duck,* and Linda Nash's *Inescapable Ecologies.*

Above all, love and thanks to my husband, Jamie, who, although he is a leg man, gives me immeasurable support, forbearance, and a steady home port; and to our little hominins, Ben and Annabel, constant reminders of the miracle of life.

NOTES

Introduction. Planet Breast

they are bigger than ever: Susan Nethero, aka "the Bra Whisperer," founder and owner, Intimacy Management Co. LLC, author interview, July 2011.

Its incidence has almost doubled: Barry A. Miller et al., "Recent Incidence Trends for Breast Cancer in Women and the Relevance of Early Detection," *CA: A Cancer Journal for Clinicians,* vol. 43 (1993), pp. 27–41. See also Stephanie E. King et al., "The 'Epidemic' of Breast Cancer in the U.S. — Determining the Factors," *Oncology,* vol. 10, no. 4 (1996), pp. 453–462.

"I would sit in the bathtub": Nora Ephron, "A Few Words about Breasts," *Esquire* (1972), republished in *Crazy Salad: Some Things about Women* (New York: Knopf, 1975), p. 4.

a piece published in the *New York Times*

Magazine: Florence Williams, "Toxic Breast Milk?" *New York Times Magazine,* January 9, 2005.

Linnaeus could have classified us: Carolus Linnaeus, *Systema Naturae,* 10th ed. (Stockholm: Laurentius Salvius, 1758).

Londa Schiebinger argues: Londa Schiebinger, *Nature's Body: Gender in the Making of Modern Science* (Boston: Beacon Press, 1993), p. 67.

"The primary biological function of breasts": Dave Barry, "Men, Get Braced; Wonderbra Coming," *Aitken Standard* (syndicated column), February 27, 1994. The rest of Barry's joke is worth repeating: "This was proved in a famous 1978 laboratory experiment wherein a team of leading male psychological researchers at Yale deliberately looked at photographs of breasts every day for two years, at the end of which they concluded that they had failed to take any notes."

Before advanced organisms produced their own estrogen: Kenneth Korach at the National Institutes of Health and Michael Baker at the University of California, San Diego, among others, posited this theory. Baker thinks our estrogen receptors retain ancient wiring once used for picking up plant, fungal, or other environmental estrogens (author interview, March 2011). Korach believes these early estrogens were critical for influencing and controlling reproduction (author interview, March 2011).

In times of trouble and stress, it may be these women: Elizabeth Cashdan, professor of anthropology, University of Utah, author interview, October 2009. Cashdan told me, "I was just sitting in a conference and there's talk after talk about what men prefer in women's body types. I got tired of it." See also Cashdan (n.d.), "Waist-to-Hip Ratio across Cultures: Trade-Offs between Androgen- and Estrogen-Dependent Traits," *Current Anthropology,* vol. 49, no. 6 (2008), pp. 1099–1107.

Chapter 1. For Whom the Bells Toll

"A 41-inch bust and a lot of perseverance": Jayne Mansfield, quoted in Raymond Strait, *Here They Are* (New York: SPI Books, 1992), p. 11.

"[Breasts] are a body part": Francine Prose, quoted in Sarah Boxer, "As a Gauge of Social Change, Behold: The Breast," *New York Times,* May 22, 1999.

"This treatment made them smooth": Mae West, *Goodness Had Nothing to Do with It* (Englewood Cliffs, N.J.: Prentice-

Hall, 1959), p. 56.

no other mammal has "breasts" the way we do: Owen Lovejoy, professor of anthropology, Kent State University, author interview, July 2010; see also R. V. Short, "The Origins of Human Sexuality" (1980), in C. R. Austin and R. V. Short (eds.), *Reproduction in Mammals and Human Sexuality,* 2nd ed. (Cambridge: Cambridge University Press, 1982), pp. 1–33.

Barnaby was preparing to publish his study: Barnaby Dixson et al., "Watching the Hourglass: Eye Tracking Reveals Men's Appreciation of the Female Form," *Human Nature,* vol. 21, no. 4 (2010), pp. 355–370.

"Whenever Barny gives seminars on waist-to-hip ratios": Scientists like studying both breasts and waist-to-hip ratios (WHRs) because they're easy to measure. To get a WHR, you divide the size of the waist by the size of the hips. The WHR for Jennifer Lopez is supposedly .67, and for both Marilyn Monroe and Venus de Milo, around .70, so their waists are 70 percent of the size of their hips. Although some anthropologists have claimed the .70 ratio is universally preferred, others point out that body mass index (BMI) is a stronger indicator of both attractiveness and fitness. One study found that women with a .70 WHR and with large breasts have higher circulating levels of estradiol, and therefore might be more fertile (see Grazyna Jasienska et al., "Large Breasts and Narrow Waists Indicate High Reproductive Potential in Women," *Proceedings of the Royal Society, London,* vol. 271 (2004), pp. 1213–1217). But the study lacks ecological relevance, meaning no one has measured whether these slightly higher hormone levels actually result in more babies being born.

an eighty-pound English bulldog named Huxley: Thomas Huxley, a biologist and contemporary of Darwin, referred to himself as "Darwin's bulldog" for his fierce defense of *On the Origin of Species.*

Alan's latest book: Alan Dixson, *Sexual Selection and the Origins of Human Mating Systems* (New York: Oxford University Press, 2009).

men have relatively small testicles: Dixson, *Sexual Selection and the Origins of Human Mating Systems,* p. 38.

"there could be a profound preference among men": Barnaby is referring to work by Frank Marlowe, "The Nubility Hypothesis," *Human Nature,* vol. 9, no. 3 (1998), pp. 263–271.

A few years ago in Brittany, France: Nicolas Gueguen, "Women's Bust Size and Men's Courtship Solicitation," *Body Image,* vol. 4 (2007), pp. 386–390.

In a similar experiment, Miss Elasto-chest tried hitchhiking: Nicolas Gueguen, "Bust Size and Hitchhiking: A Field Study," *Perceptual and Motor Skills,* vol. 105, no. 4 (2007), pp. 1294–1298.

Another study showed that waitresses with larger breasts: Michael Lynn, "Determinants and Consequences of Female Attractiveness and Sexiness: Realistic Tests with Restaurant Waitresses," *Archives of Sexual Behavior,* vol. 38, no. 5 (2009), pp. 737–745.

In his earlier data from the eye-tracker: Barnaby Dixson, Gina Grimshaw, Wayne Linklater, and Alan Dixson, "Eye-Tracking of Men's Preferences for Waist-to-Hip Ratio and Breast Size of Women," *Archives of Sexual Behavior,* vol. 40, no. 1 (2009), pp. 43–50.

Other studies have shown: Clellan Ford and Frank Beach, *Patterns of Sexual Behavior* (New York: Harper & Row, 1951), p. 88.

One study found that Western men prefer curvier women: Terry F. Pettijohn et al., "Playboy Playmate Curves:

Changes in Facial and Body Feature Preferences across Social and Economic Conditions," *Personality and Social Psychology Bulletin,* vol. 30, no. 9 (2004), pp. 1186–1197.

Barnaby expected men to prefer: For Barnaby's papers on male preferences, breast size, and areolar pigment and size, see Barnaby Dixson et al., "Men's Preferences for Women's Breast Morphology in New Zealand and Papua New Guinea," *Archives of Sexual Behavior,* (2010), e-publication ahead of print edition, available at http://www.ncbi.nlm.nih.gov/pubmed/20862533; Dixson et al., "Eye Tracking of Men's Preferences for Female Breast Size and Areola Pigmentation," *Archives of Sexual Behavior,* vol. 40, no. 1 (2011), pp. 51–58; Dixson et al., "Eye-Tracking of Men's Preferences for Waist-to-Hip Ratio and Breast Size of Women," *Archives of Sexual Behavior,* vol. 40, no. 1 (2011), pp. 43–50; Dixson et al., "Watching the Hourglass," *Human Nature,* vol. 21, no. 4 (2010), pp. 355–370.

Desmond Morris published his famous and influential book: See Desmond Morris, *The Naked Ape: A Zoologist's Study of the Human Animal* (New York: McGraw-Hill, 1967); quote from p. 67.

Elaine Morgan, a Welsh writer: For a lively read, see Elaine Morgan, *The Descent of Woman* (New York: Bantam Books, 1972); quote from p. 5.

breasts helped increase a woman's fat reserves: Typically 43.6 percent of the female physique is composed of fat in comparison to 28.4 percent in men, according to J. P. Clarys et al., "Gross Tissue Weights in the Human Body by Cadaver Dissection," *Human Biology,* vol. 56 (1984), pp. 459–473. Boguslow Pawloski also defends the idea of fat, including breast fat, as being adaptive to the woman. See Pawloski, "Center of Body Mass and the Evolution of Female Body Shape," *American Journal of Human Biology,* vol. 15, no. 2 (2003), pp. 144–150.

SWAG: I am indebted to Joseph H. Williams, professor of evolutionary biology, University of Tennessee, and a most outstanding brother-in-law, for this term.

One desert zoologist sees in breasts the camel's hump: See Ron Arieli, "Breasts, Buttocks, and the Camel Hump," *Israel Journal of Zoology,* vol. 50 (2004), pp. 87–91.

"The reasons why the breasts of women": Henri de Mondeville, quoted in Marilyn Yalom, *A History of the Breast*

(New York: Random House, 1997), p. 211.

In 1840, one physician speculated: Sir Astley Paston Cooper, *On the Anatomy of the Breast* (London: Longman, Orme, Green, Brown, and Longman's, 1840), p. 59.

an Israeli researcher posited that fatty breasts: Arieli, "Breasts, Buttocks, and the Camel Hump."

"ensures that the nipple is no longer anchored": Elaine Morgan, *The Descent of the Child* (New York: Oxford University Press, 1995), p. 47.

We may be the only mammal: Daniel Lieberman, professor of human evolutionary biology, Harvard University, author interview, August 2011. I will note that Lieberman warned me away from making too much of the basicranial flexion argument. Just as it is difficult to know when pendulous breasts evolved, it is also difficult to know when speech evolved or how closely speech, neck, and breasts may be related. Point taken.

"They're pretty, they're flamboyant": Natalie Angier, *Woman: An Intimate Geography* (New York: Random House, 1999), p. 124.

Chapter 2. Circular Beginnings

". . . from so simple a beginning": Charles Darwin, *On the Origin of Species* (New York: Thomas Y. Crowell, 1860), p. 460.

The manatee has nipples under her flippers: The information on mammal features came from various sources, including Olav Oftedal, author interview, March 2010; Alan Dixson, author interview, June 2010; Sandra Steingraber, *Having Faith: An Ecologist's Journey to Motherhood* (Cambridge, Mass.: Perseus, 2001), p. 215; and, on the opossum, "With the Wild Things," at http://digitalcollections.fiu.edu/wild/transcripts/possums1.htm (accessed October 2011).

the ability to lactate is among our most valuable genetic assets: Bruce German, professor of food science and technology, University of California, Davis, author interview, October 2010.

one-sixth the protein found: On milk fat compositions of various species, see Caroline Pond, "Physiological and Ecological Importance of Energy Storage," Symposia of the Zoological Society of London, *Physiological Strategies in Lactation,* vol. 51 (1984), pp. 1–29.

The earliest lactating species: Sarah

Blaffer Hrdy, *Mothers and Others: The Evolutionary Origins of Mutual Understanding* (Cambridge, Mass.: Belknap Press, 2009), p. 39; and M. Peaker, "The Mammary Gland in Mammalian Evolution: A Brief Commentary on Some of the Concepts," *Journal of Mammary Gland Biology and Neoplasia,* vol. 7, no. 3 (2002), p. 347.

Mammals owned the Cenozoic: For readable discussions of the ascendance of mammals, see T. S. Kemp, *The Origin and Evolution of Mammals* (New York: Oxford University Press, 2005); and Donald R. Prothero, *After the Dinosaurs: The Age of Mammals* (Bloomington: Indiana University Press, 2006).

Darwin himself went out on a limb: Discussed in Charles Darwin, *On the Origin of Species* (New York: Penguin, 2009; first published 1859), pp. 322–323.

we would never have breasts if we didn't have teeth: Neil Shubin, *Your Inner Fish: A Journey into the 3.5-Billion-Year History of the Human Body* (New York: Random House, 2008), p. 78.

even what sex the fetus is in order to fine-tune the composition of the milk: Katherine Hinde, assistant professor, Department of Human Evolutionary Biology, Harvard University, author interview, December 2010.

The first fluid was a sort of natural Lysol: There are a number of fascinating journal articles about the origins of the mammary gland and its beginnings as part of the innate immune system. I recommend D. G. Blackburn et al., "The Origins of Lactation and the Evolution of Milk: A Review with New Hypotheses," *Mammal Review,* vol. 19 (1989), pp. 1–26; D. G. Blackburn, "Evolutionary Origins of the Mammary Gland," *Mammal Review,* vol. 21 (1991), pp. 81–96; and two Oftedal papers: "The Mammary Gland and Its Origin during Synapsid Evolution," *Journal of Mammary Gland Biology and Neoplasia,* vol. 7, no. 3 (July 2002), pp. 225–252; and "The Origin of Lactation as a Water Source for Parchment-Shelled Eggs," *Journal of Mammary Gland Biology and Neoplasia,* vol. 7, no. 3 (July 2002), pp. 253–266.

Lactation, with its tremendous metabolic efficiencies: Kemp, *Origin and Evolution of Mammals,* p. 113.

Chapter 3. Plumbing

"I have heard a good anatomist say": Astley Paston Cooper, *On the Anatomy of the Breast* (London: Lea & Blanchard,

1845), p. 6.

Napoleon's penis: See Tony Perrottet, *Napoleon's Privates: 2,500 years of History Unzipped* (New York: HarperCollins, 2008), pp. 20–27; and Charles Hamilton, *Auction Madness: An Uncensored Look behind the Velvet Drapes of the Great Auction Houses* (New York: Everest House, 1981), pp. 54–55.

woman said to have the largest implants in the world: Fox News reported that the Houston woman, Sheyla Hershey, suffered a serious staph infection after her latest implant surgery. It was her thirtieth operation, according to "Woman with World's Largest Breasts Fighting for Her Life," July 14, 2010, available at http://www.foxnews.com/health/2010/07/14/woman-worlds-largest-breasts-fighting-life/#ixzz1DmS8FrTD.

"What goes up must go down": Patrick McCain, "World's Largest Breasts, 38KKK Sheyla Hershey Breast Implants Removed," Rightpundits.com, September 14, 2010 (originally published in 2009), at http://www.rightpundits.com/?p=2822.

Over the course of a menstrual cycle: Z. Hussain et al., "Estimation of Breast Volume and Its Variation during the Menstrual Cycle Using MRI and Stereology,"

British Journal of Radiology, vol. 72, no. 855 (1999), pp. 236–245.

"Brassiere design is one engineering activity": Quote and equation from Edward Nanas, "Brassieres: An Engineering Miracle," *Science and Mechanics,* February 1964, available at http://www .firstpr.com.au/show-and-tell/corsetry-1/ *nanas*/engineer.html (accessed October 2011). Nanas backed up his statement with a description from Mrs. Ida Rosenthal, the seventy-seven-year-old head of Maidenform. "She recently returned from a tour of the Soviet garment industry and found that bra designers on the other side of the Iron Curtain have not yet discovered stretch fabrics, foam padding, hooks and eyes, or the strapless bra."

She showed me the action footage: To see Werb's film clips, check out http:// anatomy.ucsf.edu/Werbwebsite/egebald %20movies%202008/Movie_1.mov.

a digression: On the cadaver trade, see Julie Bess Frank, "Body Snatching: A Grave Medical Problem," *Yale Journal of Biology and Medicine,* vol. 49 (1976), pp. 399–410; and W. B. Walker, "Medical Education in 19th Century Great Britain," *Journal of Medical Education,* vol. 31, no. 11 (1956), pp. 765–777.

"a breadth of experience unparalleled before or since": James Going, clinical senior lecturer in pathology, University of Glasgow, author interview, May 2010.

"galactograms": Cooper, *On the Anatomy of the Breast;* for a digital version, see http://jdc.jefferson.edu/cooper/61/.

on rare occasion have been able to produce milk-like fluid: For more on male lactation, see Jared Diamond, who lays out a plausible male breast-feeding scenario in "Father's Milk," *Discover,* vol. 16, no. 2 (February 1995), pp. 82–87. This essay perhaps inspired a Swedish college student named Ragnar "Milkman" Bengtsson, who, in 2009, tried to stimulate milk production by pumping his nipples every three hours for two months. It didn't work. See "Swedish 'Milkman' Loses Breastfeeding Battle," *The Local,* December 1, 2009, at http://www.thelocal.se/ 23592/20091201/. The anthropologist Barry Hewlett documented suckling among men of the Aka Pygmy tribe in central Africa, but they appeared to be providing "comfort suckling" and not nutrition. See Joanna Moorhead, "Are the Men of the African Aka Tribe the Best Fathers in the World?" *The Guardian,* July

15, 2005.

"the breasts are generally two in number:" Cooper, *On the Anatomy of the Breast,* p. 13.

Chapter 4. Fill Her Up

". . . but on the fourth night": Maria Edgeworth, *Tales and Novels: Harrington; Thoughts on Bores; Ormond* (London: George Routledge and Sons, 1893), p. 394.

cosmetic surgery: Statistics are from the American Society for Aesthetic and Plastic Surgery, "Statistics," Press Center, at http://www.surgery.org/media/statistics (accessed October 2011).

performs more augmentations by far than any doctor in Texas: Becca Quisenberry, Patient Coordinator, Ciaravino Plastic Surgery, author interview, September 2011.

Falsies, made out of wire, sheet metal, papier-mâché: See Teresa Riordan, "We Must Increase Our Bust: A History of Breast Enhancement, Told in Patent Drawings," *Slate,* April 11, 2005, at http://www.slate.com/id/2116481; and Elizabeth Haiken, *Venus Envy: A History of Cosmetic Surgery* (Baltimore: Johns Hopkins University Press, 1997), pp. 243–246.

Consider the case of poor Elisabeth Trevers: Gordon Letterman and Maxine Schurter, "Will Durston's Mammaplasty," *Plastic and Reconstructive Surgery,* vol. 53, no. 1 (1974), quoted in Nora Jacobsen, *Cleavage: Technology, Controversy, and the Ironies of the Man-Made Breast* (New Brunswick, N.J.: Rutgers University Press, 2000), p. 50.

Vincenz Czerny: The first boob job was technically a reconstruction. See Theodore W. Uroskie Jr. and Lawrence B. Colen, "History of Breast Reconstruction," *Seminars in Plastic Surgery,* vol. 18, no. 2 (May 2004), pp. 65–69.

glass balls, ivory, wood chips: For information on the early-twentieth-century implant materials used, see Haiken, *Venus Envy;* also Bernard M. Patten, former chief of neuromuscular disease, Baylor College of Medicine, author interview, February 2011.

disadvantages of paraffin: Jacobsen, *Cleavage,* pp. 52–54.

Women have painted their faces: Julie M. Spanbauer, "Breast Implants as Beauty Ritual: Woman's Sceptre and Prison," *Yale Journal of Law and Feminism,* vol. 9, no. 157 (1997).

Ivalon: S. Murthy Tadavarthy, James H.

Moller, and Kurt Amplatz, "Polyvinylalcohol (Ivalon) — A New Embolic Material," *American Journal of Roentgenology,* vol. 125, no. 3 (November 1975), pp. 609–616.

"The material's one drawback": Plastic surgeon Milton Edgerton, paraphrased in *Jet Magazine,* December 12, 1957, available at http://www.flickr.com/photos/vieilles_annonces/3778246964/ (accessed October 2011).

Dow Corning: For the history of Dow Corning, see Haiken, *Venus Envy,* pp. 246–247. Also see this colorful document from the Dow website: www.dowcorning.com/content/publishedlit/01-4027-01.pdf (accessed October 2011).

the breasts of Japanese prostitutes, who were being injected with it: M. Sharon Webb, "Cleopatra's Needle: The History and Legacy of Silicone Injections," Harvard Law School paper, January 1997, available at http://leda.law.harvard.edu/leda/data/197/mwebb.pdf; and Haiken, *Venus Envy,* p. 246.

Back in Houston, plastic surgeon Thomas Cronin: As recounted by Thomas Biggs, retired plastic surgeon, Houston, Texas, author interview, January 2011. Biggs also recounted the story of

Cronin's ambition and the first surgery, including parts about Esmerelda. He was a resident of Cronin's at the time.

On the back of the bag, they added several patches of Dacron: Jacobsen, *Cleavage,* pp. 78–79.

Some authors say they tested it in six dogs: John Byrne, *Informed Consent* (New York: McGraw-Hill, 1997), pp. 47–50.

Surgery to remove implants, known as explantation: For a discussion of what's found at the explant site, see R. Vaamonde et al., "Silicone Granulomatous Lymphadenopathy and Siliconomas of the Breast," *Histology and Histopathology,* vol. 4 (October 1997), pp. 1003–1011.

"Many women with limited development of the breast": Gerow, quoted in Jacobsen, *Cleavage,* pp. 78–79.

One surgeon's autobiography: Robert Alan Franklyn, *Beauty Surgeon* (Long Beach, Calif.: Whitehorn, 1960).

"there is a substantial and enlarging body": The American Society of Plastic and Reconstructive Surgery's statement to the Food and Drug Administration, in 1982, is a well-known quote. I love the not-quite-subliminal use of the word *enlarging.* See the quote referenced with biting commentary from Barbara Ehrenreich, "Stamping Out a Dread Scourge," *Time,* February 17, 1992, available at http://www.time.com/time/magazine/article/0,9171,974902,00.html.

The largest was called "the Burlesque": Jacobsen, *Cleavage,* p. 79.

Gerow reputedly liked big breasts: Bernard Patten, author interview, January 2011.

One Houston doctor boasted: For an excellent article on Houston in its boob-job glory days, see Mimi Swartz, "Silicone City," *Texas Monthly,* vol. 23, no. 8 (1995), pp. 64–78.

By 1985, one hundred thousand women: *Newsweek* noted in 1985 that nearly one hundred thousand breast augmentations had been performed over the past year "for a total addition to the nation's mammary capacity of some 13,000 gallons (of silicone gel)," as quoted in Haiken, *Venus Envy,* p. 273.

Carol Doda: For more on Doda, there's a great section on her breasts (and the infamous piano) in Mike Sinclair's *San Francisco: A Cultural and Literary History* (Oxford: Signal Books, 2004), pp. 84–85.

"Carol Doda's breasts are up there": Thomas Wolfe, *The Pump House Gang* (New York: Bantam, 1969), p. 67.

strippers instantly saw their tips increase: Bernard Patten, author interview, January 2011.

Houston became the strip-club capital of the world: Michael Ciaravino, author interview, January 2011.

Rick's Cabaret: For more on the history of Rick's Cabaret, see Swartz, "Silicone City."

restricted the use of the "medical-grade" stuff: On the FDA restrictions on the "medical-grade" stuff and reports of infection, gangrene, and so on, see Haiken, *Venus Envy,* pp. 274–275.

Many patients — 41 percent, according to a 1979 study: G. P. Hetter, "Satisfactions and Dissatisfactions in Patients with Augmentation Mammaplasty," *Plastic and Reconstructive Surgery,* vol. 64, no. 2 (August 1979), p. 151.

An enormous percentage of patients — around 25 to 70 percent: Neal Handel et al., "A Long-Term Study of Outcomes, Complications, and Patient Satisfaction with Breast Implants," *Plastic and Reconstructive Surgery,* vol. 117, no. 3 (March 2006), pp. 757–767.

"They were like doorbells": Bernard Patten, author interview, January 2011.

Early dissections of the affected tissue:

Thomas Biggs, author interview, January 2011.

The operation itself: Michael Ciaravino, author interview, January 2011.

Company salesmen were told to wash the leaking implants: This comes from an internal Dow Corning memo dated January 15, 1975, that was made public when the group Public Citizen sued the FDA; cited in Jack Doyle, *Trespass against Us: Dow Chemical and the Toxic Century* (Monroe, Maine: Common Courage Press, 2004), p. 257.

causing a prolonged inflammatory response and "microencapsulations": See Michelle Copeland et al., "Absent Silicone Shell in a MEME Polyurethane Silicone Breast Implant: Report of a Case and Review of the Literature," *Breast Journal,* vol. 2, no. 5 (September 1996), pp. 340–344.

manufacturer of the foam was apparently surprised: See Nicholas Regush, "Toxic Breasts," *Ms. Magazine,* vol. 17, no. 1 (January/February 1992), pp. 24–31.

Many surgeons remember these implants fondly: Thomas Biggs, author interview, January 2011.

foam-covered implants continue to be used: Handel et al., "Long-Term Study of Outcomes."

"we know more about the life span of automobile tires": David Kessler, quoted in Spanbauer, "Breast Implants as Beauty Ritual."

Dow Corning declared bankruptcy: On Dow Corning's bankruptcy history, see Dow Corning's publication "Highlights from the History of Dow Corning Corporation, the Silicone Pioneer," available at www.dowcorning.com/content/published lit/01-4027-01.pdf (accessed October 10, 2011); and John Schwartz, "Dow Corning Accepts Implant Settlement Plan; $3.2 Billion Earmarked for Health Claims," *Washington Post,* July 9, 1998.

anaplastic large-cell lymphoma: Denise Grady, "Breast Implants Are Linked to Rare but Treatable Cancer, F.D.A. Finds," *New York Times,* January 26, 2011.

"The message that never reaches the public": Spanbauer, "Breast Implants as Beauty Ritual."

from a high of 150,000: Marcia Angell, "Breast Implants — Protection or Paternalism?" *New England Journal of Medicine,* vol. 326 (June 18, 1992), pp. 1695–1696.

the FDA approved: On FDA approvals, see "Breast Implants," U.S. Food and Drug Administration, at http://www.fda

.gov/MedicalDevices/ProductsandMedi
calProcedures/ImplantsandProsthetics/
BreastImplants/default.htm (accessed
October 14, 2011).

roughly $820 million a year: Denise
Grady, "Dispute over Cancer Tied to
Implants," *New York Times,* February 17,
2011.

Between five and ten million women:
Grady, "Dispute over Cancer Tied to Implants."

product insert data sheet: For Mentor's
product insert data sheet, see http://www
.mentorwwllc.com/global-us/SafetyInfor
mation.aspx (accessed October 2011).

**a major review of the literature from
the Institute of Medicine:** *Safety of
Silicone Breast Implants* (Washington,
D.C.: Institute of Medicine National
Academy Press, 2000). Also available
through the Institute of Medicine website,
at www.iom.edu.

**"It is not known if a small amount of
silicone":** For information on the effects
on nursing infants, see "FDA Breast
Implant Consumer Handbook — 2004,"
at http://www.fda.gov/MedicalDevices/
ProductsandMedicalProcedures/Implants
andProsthetics/BreastImplants/ucm064
106.htm.

"Fortunately, patients undergoing plastic surgery of the breast": Eugene H. Courtiss and Robert M. Goldwyn, "Breast Sensation before and after Plastic Surgery," *Plastic and Reconstructive Surgery,* vol. 58, no. 1 (July 1976), pp. 1–13, quoted in Haiken, *Venus Envy,* p. 270.

Mentor's study to date: Summary of the study is available at http://www.mentor wwllc.com/global-us/SafetyInformation .aspx (accessed October 2011).

critics like Naomi Wolf: Naomi Wolf, *The Beauty Myth* (New York: William Morrow, 1991), p. 242.

Barbie's proportions are naturally found: Kevin I. Norton et al., "Ken and Barbie at Life Size," *Sex Roles,* vol. 34, no. 3–4 (1996), pp. 287–294.

a twenty-nine-year-old named Gloria: I've changed the names of Dr. C's patients to protect their privacy.

Chapter 5. Toxic Assets

"I tell people I come from a different planet": Sylvia Earle, author interview, February 10, 2012.

nature writer and biologist had received a disturbing letter: For a description of Huckins's experience with DDT, see Eleni Himaras, "Rachel Carson's Groundbreak-

ing 'Silent Spring' Was Inspired by Duxbury Woman," *Patriot Ledger (Quincy,
Mass.)*, May 26, 2007.

"elixirs of death": Rachel Carson, *Silent
Spring* (New York: Ballantine Books,
1962), pp. 24–43.

**"The sedge is wither'd from the lake, /
And no birds sing":** Carson, *Silent
Spring,* p. 12.

**"For the first time in the history of the
world":** Carson, *Silent Spring,* p. 25.

**"I was born with a plastic spoon in my
mouth":** The Who, "Substitute," 1966.

The term *endocrine disruptor*: Theo Colborn, founder and president of The Endocrine Disruption Exchange and professor
emeritus of zoology, University of Florida,
Gainesville, author interview, March 2010.

**women reportedly get better at verbal
and fine-motor skills:** For example, see
Elizabeth Hampson, "Estrogen-Related
Variations in Human Spatial and
Articulatory-Motor Skills," *Psychoneuro-
endocrinology,* vol. 15, no. 2 (1990), pp.
97–111.

do humans cultivate marijuana: For
more on the wonders of evolutionary
adaptations of marijuana, see Michael Pol-
lan, *The Botany of Desire: A Plant's-Eye*

View of the World (New York: Random House, 2001).

Giant fennel, found by the Greeks in the seventh century BC: Timothy Taylor, *The Prehistory of Sex: Four Million Years of Human Sexual Culture* (New York: Bantam Books, 1996), p. 90.

BPA's molecular structure is simple and elegant: Jeffrey Stansbury, polymer chemist, University of Colorado, Denver, author interview, March 2011.

Now produced in mind-boggling quantities: Fact sheet, "Bisphenol A (BPA) and Breast Cancer," published by the Breast Cancer Fund, December 8, 2008, available at www.breastcancerfund.org/assets/pdfs/bpaandbc_factsheet_120808 .pdf.

DES: For the effects of DES on daughters and sons, see Nancy Langston, *Toxic Bodies: Hormone Disruptors and the Legacy of DES* (New Haven, Conn.: Yale University Press, 2010), p. 135.

BPA has been shown to cause: A. G. Recchia et al., "Xenoestrogens and the Induction of Proliferative Effects in Breast Cancer Cells via Direct Activation of Oestrogen Receptor Alpha," *Food Additives and Contaminants,* vol. 21 (2004), pp. 134–144; S. V. Fernandez and J. Russo,

"Estrogen and Xenoestrogens in Breast Cancer," *Toxicologic Pathology,* vol. 38, no. 1 (January 2010), pp. 110–122.

There is something about BPA: Sarah Jenkins et al., "Oral Exposure to Bisphenol A Increases Dimethylbenzanthracene-Induced Mammary Cancer in Rats," *Environmental Health Perspectives,* vol. 117, no. 6 (June 2009), pp. 910–915.

In other rat experiments: Milena Durando et al., "Prenatal Bisphenol A Exposure Induces Preneoplastic Lesions in the Mammary Gland in Wistar Rats," *Environmental Health Perspectives,* vol. 115, no. 1 (January 2007), pp. 80–86.

Higher EZH2 levels are associated with an increased risk: Leo F. Doherty et al., "In Utero Exposure to Diethylstilbestrol (DES) or Bisphenol-A (BPA) Increases EZH2 Expression in the Mammary Gland: An Epigenetic Mechanism Linking Endocrine Disruptors to Breast Cancer," *Hormones and Cancer,* vol. 1, no. 3 (2010), pp. 146–155.

Scientists call this "phenotypic plasticity": For an interesting overview, see Richard G. Bribiescas and Michael P. Muehlenbein, "Evolutionary Endocrinology," in Michael P. Muehlenbein (ed.), *Human Evolutionary Biology* (New York:

Cambridge University Press, 2010), pp. 127, 137.

DES was still manufactured: Furthermore, its illegal use as a growth hormone in cattle continued well into the 1980s. For more on DES and its dates of use, see http://www.websters-online-dictionary.org/definitions/Diethylstilbestrol; Nancy Langston offers a compelling history in *Toxic Bodies,* p. 117; see also Orville Schell, *Modern Meat: Antibiotics, Hormones, and the Pharmaceutical Farm* (New York: Vintage, 1985), p. 331.

In the United States, every chemical is assumed safe: Lynn Goldman, "Preventing Pollution? U.S. Toxic Chemicals and Pesticides Policies and Sustainable Development," *Environmental Law Reporter,* vol. 32 (2002), pp. 11018–11041.

Of the 650 top-volume chemicals in use: Rick Smith and Bruce Lourie, *Slow Death by Rubber Duck* (Berkeley, Calif.: Counterpoint, 2009), p. xiv.

"They leave the mammary gland in the trash can": Ruthann Rudel, director of research, Silent Spring Institute, author interview, February 2011. See also Ruthann Rudel et al., "Mammary Gland Development as a Sensitive Indicator of Early Life Exposures: Recommendations from an Interdisciplinary Workshop," presented at The Mammary Gland Evaluation and Risk Assessment Workshop in Oakland, Calif., November 2009. Also see S. L. Makris, "Current Assessment of the Effects of Environmental Chemicals on the Mammary Gland in Guideline EPA, OECD, and NTP Rodent Studies," *Environmental Health Perspectives,* vol. 119, no. 8 (2011), pp. 1047–1052; and Florence Williams, "Scientists to Chemical Regulators: Stop Ignoring Boobs," *Slate,* June 27, 2011, available at http://www .slate.com/articles/double_x/doublex/2011/ 06/scientists_to_chemical_regulators_stop _ignoring_boobs.single.html#comments.

In the body it appears to increase: J. L. Raynor et al., "Adverse Effects of Prenatal Exposure to Atrazine during a Critical Period of Mammary Gland Growth," *Journal of Toxicological Sciences,* vol. 87 (2005), pp. 255–266.

The journal *Cancer* reported in 2007: Ruthann A. Rudel et al. "Chemicals Causing Mammary Gland Tumors in Animals Signal New Directions for Epidemiology, Chemicals Testing, and Risk Assessment for Breast Cancer Prevention," *Cancer,* vol. 109, no. 12 (2007, Supplement), pp. 2635–2666.

Roughly one thousand chemicals: Theo Colborn, author interview, March 2010.

"The possibilities of DDT are sufficient": For this quotation from Simmons and other information on DDT, see Will Allen, *The War on Bugs* (White River Junction, Vt.: Chelsea Green, 2008), p. 171.

by the early 1970s, 1.3 trillion pounds had been sprinkled: EPA report, "DDT Regulatory History: A Brief Survey (to 1975)," excerpted from *DDT, A Review of Scientific and Economic Aspects of the Decision to Ban Its Use as a Pesticide,* prepared for the Committee on Appropriations of the U.S. House of Representatives by EPA, July 1975, available at http://www.epa.gov/history/topics/ddt/02.htm.

shortened duration of lactation: For a good introduction to the potential links between chemicals and mammary gland dysfunction, see Ruthann Rudel et al., "Environmental Exposures and Mammary Gland Development: State of the Science, Public Health Implications, and Research Recommendations," *Environmental Health Perspectives,* vol. 119, no. 8 (August 2011), pp. 1053–1061; also available at http://ehp03.niehs.nih.gov/article/fetchArticle.action?articleURI=info%3Adoi%2F

10.1289%2Fehp.1002864.

The younger women, the ones exposed to the most DDT: See Barbara A. Cohn et al., "DDT and Breast Cancer in Young Women: New Data on the Significance of Age at Exposure," *Environmental Health Perspectives,* vol. 115, no. 10 (October 2007), pp. 1406–1414.

women born after 1940 have much higher levels: See Tom Reynolds, "Study Clarifies Risk of Breast, Ovarian Cancer among Mutation Carriers," *Journal of National Cancer Institute,* vol. 95, no. 24 (2003), pp. 1816–1818.

Cheap by-products of fossil-fuel production: Theo Colborn, "Foreword," in Smith and Lourie, *Slow Death by Rubber Duck,* pp. viii–x.

Today we use thirty times more synthetic pesticides: Theo Colborn, Diane Dumanoski, and John Peterson Myers, *Our Stolen Future: Are We Threatening Our Fertility, Intelligence, and Survival? — A Scientific Detective Story* (New York: Penguin Books, 1996), p. 138.

Now, the rate is 1 out of 2.5: "Lifetime Risk of Developing or Dying from Cancer," from the *American Cancer Society,* available at http://www.cancer.org/Cancer/CancerBasics/lifetime-probability-of-developing-or-dying-from-cancer. **Chemical World News reacted:** James Stuart Olson, *Bathsheba's Breast: Woman, Cancer and History* (Baltimore: Johns Hopkins University Press, 2002), p. 226.

"If we are going to live so intimately with these chemicals": Carson, *Silent Spring,* p. 17.

Chapter 6. Shampoo, Macaroni, and the American Girl

"Still she went on growing": Lewis Carroll, *Alice's Adventures in Wonderland and Through the Looking Glass* (New York: Macmillan Co., 1897), p. 45.

ship them to Canada for testing: We used Axys Analytical Services, Sidney, British Columbia.

girls were developing breasts and sprouting pubic hair: Marcia E. Herman-Giddens et al., "Secondary Sexual Characteristics and Menses in Young Girls Seen in Office Practice: A Study from the Pediatrics in Office Settings (PROS) Network, American Academy of Pediatrics," *Pediatrics,* vol. 99, no. 4 (1997), pp. 505–512.

2007 report for the Breast Cancer Fund: Sandra Steingraber, "The Falling Age of Puberty in U.S. Girls: What We Know,

What We Need to Know," published by the Breast Cancer Fund (2007), p. 24, available at http://www.breastcancerfund.org/assets/pdfs/publications/falling-age-of-puberty.pdf.

"We think that puberty": Suzanne Fenton, research biologist, Reproductive Endocrinology Group, National Institute of Environmental Health Sciences, author interview, December 2007.

If you get your first period before age twelve: Steingraber, "Falling Age of Puberty in U.S. Girls," p. 24.

by 2011, one-third of black girls: F. M. Biro et al., "Pubertal Assessment Method and Baseline Characteristics in a Mixed Longitudinal Study of Girls," *Pediatrics,* vol. 126, no. 3 (September 2010), pp. 583–590.

The tragic result, according to pediatrician Sharon Cooper: Patricia Leigh Brown, "In Oakland, Redefining Sex Trade Workers as Abuse Victims," *New York Times,* May 23, 2011.

the age of sexual maturity in girls has dropped slowly but steadily: Anne-Simone Parent et al., "The Timing of Normal Puberty and the Age Limits of Sexual Precocity: Variations around the World, Secular Trends, and Changes after

Migration," *Endocrine Reviews,* vol. 24, no. 5 (2003), pp. 668–693.

a whopping 13 million calories: Sarah Blaffer Hrdy, *Mothers and Others: The Evolutionary Origins of Mutual Understanding* (Cambridge, Mass.: Belknap Press, 2009), p. 31.

When poor, once-hungry immigrants: Peter D. Gluckman and Mark A. Hanson, "Evolution, Development and Timing of Puberty," *Trends in Endocrinology and Metabolism,* vol. 17, no. 1 (2006), pp. 7–12.

Adopted Indian girls who move to Sweden as infants: Parent et al., "Timing of Normal Puberty."

"For the first time in our evolutionary history": Peter Gluckman, professor of paediatric and perinatal biology, University of Auckland, author interview, January 2010.

"chemical, physical and social factors interact with genes": As stated in a public talk by Robert Hiatt, Department of Epidemiology and Biostatistics, University of California, San Francisco, and principal investigator at Breast Cancer and the Environment Research Centers (BCERC), Cavallo Point, California, November 2011.

the percentage of American girls aged

six to eleven: "Health, United States, 2008, with Special Feature on the Health of Young Adults," Centers for Disease Control and Prevention, National Center for Health Statistics, February 18, 2009. For highlights of the report, see http://www.cdc.gov/nchs/pressroom/09newsreleases/hus08.htm.

Fat has been called "the third ovary": Debbie Clegg, assistant professor of internal medicine, University of Texas Southwestern Medical Center, author interview, November 2008.

girls who reached puberty earlier ate more meat: Imogen S. Rogers et al., "Diet throughout Childhood and Age at Menarche in a Contemporary Cohort of British Girls," *Public Health Nutrition,* vol. 13, no. 12 (2010), pp. 2052–2063.

"The nutritional factor consistently associated with timing of puberty": Frank Biro, director of adolescent medicine, Cincinnati Children's Hospital, author interview, July 2009.

Pubic hair is influenced more by adrenal hormones: Parent et al., "Timing of Normal Puberty," p. 668.

In a recent Swedish study: Jingmei Li et al., "Effects of Childhood Body Size on Breast Cancer Tumour Characteristics,"

Breast Cancer Research, vol. 12, no. 2 (2010), pp. 1–9.

breast-feeding rates in Denmark: For breast-feeding rates, see the Centers for Disease Control and Prevention, Breast-feeding Report Card (2010), at http://www.cdc.gov/breastfeeding/data/reportcard .htm.

women who work under lights on the night shift: Joe Russo and Irma Russo, *The Molecular Basis of Breast Cancer: Prevention and Treatment* (Berlin: Springer-Verlag, 2003), p. 5.

Studies have found that girls with precocious puberty: Steingraber, "Falling Age of Puberty in U.S. Girls," p. 37.

it's been documented that girls not living with a biological father: For example, see Julianna Deardorff et al., "Father Absence, Body Mass Index, and Pubertal Timing in Girls: Differential Effects by Family Income and Ethnicity," *Journal of Adolescent Health,* vol. 48, no. 5 (2011), pp. 441–447.

Female elephants do something similar: In zoos, elephants reach maturity at age eleven, compared to sixteen to eighteen in the wild. Unfortunately for them, it doesn't always translate into higher fertility; see http://buzzle.com/editorials/10-22-

2002-28715.asp. Wildlife biologists Mark and Delia Owens documented an eight-year-old female with a newborn in a region of Zambia that had been heavily hunted by ivory poachers. The mother, an orphan with no guiding adults, was "not a good mother." Recounted in Mark Owens and Delia Owens, *Secrets of the Savannah* (New York: Houghton Mifflin, 2006), p. 133.

families are actually slightly more stable: Lise Aksglaede, author interview, July 2009.

the once-rare birth defect of undescended testicles: Leonard J. Paulozzi, "International Trends in Rates of Hypospadias and Cryptorchidism," *Environmental Health Perspectives,* vol. 107, no. 4 (April 1999), pp. 297–302.

In a study of 1,600 babies born between 1997 and 2001: Katharina M. Main et al., "Larger Testes and Higher Inhibin B Levels in Finnish Than in Danish Newborn Boys," *Journal of Clinical Endocrinology and Metabolism,* vol. 91, no. 7 (2006), pp. 2732–2737. For more on male genital defects and possible environmental links, see Florence Williams, "The Little Princes of Denmark," *Slate,* February 24, 2010, available at http://www.slate.com/articles/

double_x/doublex_health/2010/02/the_
little_princes_of_denmark.html.

"In the first photo, a four-and-a-half-year-old girl": Orville Schell, *Modern Meat: Antibiotics, Hormones, and the Pharmaceutical Farm* (New York: Vintage, 1985), p. 283.

One study did find unusually high levels of phthalates: Ivelisse Colon et al., "Identification of Phthalate Esters in the Serum of Young Puerto Rican Girls with Premature Breast Development," *Environmental Health Perspectives,* vol. 108, no. 9 (September 2000), pp. 895–900.

recently linked to genital abnormalities: Shanna H. Swan et al., "Decrease in Anogenital Distance among Male Infants with Prenatal Phthalate Exposure," *Environmental Health Perspectives,* vol. 113, no. 8 (August 2005), pp. 1056–1061; S. H. Swan et al., "Prenatal Phthalate Exposure and Reduced Masculine Play in Boys," *International Journal of Andrology,* vol. 33, no. 2 (April 2010), pp. 259–269; Shanna H. Swan, "Environmental Phthalate Exposure in Relation to Reproductive Outcomes and Other Health Endpoints in Humans," *Environmental Research,* vol. 108, no. 2 (August 11, 2008), pp. 177–184.

Girls in New York City: Mary S. Wolff et al., "Investigation of Relationships between Urinary Biomarkers of Phytoestrogens, Phthalates, and Phenols and Pubertal Stages in Girls," *Environmental Health Perspectives,* vol. 118, no. 7 (July 2010), pp. 1039–1046.

When the BCERC researchers in Cincinnati: Susan Pinney, Department of Environmental Health, University of Cincinnati College of Medicine, author interview, September 2011.

A newspaper in Taiwan: Shelley Huang, "Smell May Indicate Plasticizers: Experts," *Taipei Times,* June 6, 2011.

sniffed it out in everything: Environmental Working Group, "Pesticide in Soap, Toothpaste and Breast Milk — Is It Kid-Safe?" Washington, D.C., July 17, 2008; see also Antonia M. Calafat et al., "Urinary Concentrations of Triclosan in the U.S. Population: 2003–2004," *Environmental Health Perspectives,* vol. 116, no. 3 (March 2008), pp. 303–307.

"have become ubiquitous in the environment": Dominique J. Williams, Division of Health Sciences, U.S. Consumer Product Safety Commission, "Toxicity Review of Di-n-butyl Phthalate," staff assessment memo, April 7, 2010, available at www.cpsc.gov/about/cpsia/toxicityDBP .pdf. See also Susan M. Duty et al., "The Relationship between Environmental Exposures to Phthalates and DNA Damage in Human Sperm Using the Neutral Comet Assay," *Environmental Health Perspectives,* vol. 111, no. 9 (July 2003), pp. 1164–1169; and Mary S. Wolff et al., "Pilot Study of Urinary Biomarkers of Phytoestrogens, Phthalates, and Phenols in Girls," *Environmental Health Perspectives,* vol. 115, no. 1 (January 2007), pp. 116–121. For more information about levels of these chemicals in the general U.S. population, see the Centers for Disease Control and Prevention, *Fourth National Report on Human Exposure to Environmental Chemicals,* July 2010, available at http://www.cdc.gov/exposure report/.

It has been associated with liver toxicity: See Centers for Disease Control and Prevention, "Chemical Information: Di-2-ethylhexyl Phthalate," *National Report on Human Exposure to Environment Chemicals,* November 15, 2010, available at http:// www.cdc.gov/exposurereport/data_tables/ DEHP_ChemicalInformation.html.

"The results were rather striking to us":

G. D. Bittner et al., "Most Plastic Products Release Estrogenic Chemicals: A Potential Health Problem That Can Be Solved," *Environmental Health Perspectives,* vol. 119, no. 7 (2011), pp. 989–996.

Some scientists argue that in addition to our genome: For example, see Christopher Paul Wild, "Complementing the Genome with an 'Exposome': The Outstanding Challenge of Environmental Exposure Measurement in Molecular Epidemiology," *Cancer Epidemiology, Biomarkers and Prevention,* vol. 14, no. 8 (2005), pp. 1847–1850.

atomic bombs dropped on Hiroshima and Nagasaki: On breast cancer rates among Hiroshima and Nagasaki bomb survivors, see Masayoshi Tokunaga et al., "Incidence of Female Breast Cancer among Atomic Bomb Survivors, 1950–1985," *Radiation Research,* vol. 182, no. 2 (1994), pp. 209–223.

Women with the BRCA genes: Nancy Langston, *Toxic Bodies: Hormone Disruptors and the Legacy of DES* (New Haven, Conn.: Yale University Press, 2010), p. 12.

"ecological disorder": Steingraber, "Falling Age of Puberty in U.S. Girls," p. 59.

recent experiments with adolescent rats: For research on fat and inflammation in adolescent mice, see Sandra Z. Haslam, "Is There a Link between a High-Fat Diet during Puberty and Breast Cancer Risk?" *Women's Health,* vol. 7, no. 1 (2011), pp. 1–3.

Forty-four percent of TV ads: Charles Atkins, professor, Department of Communications, Michigan State University, author interview, November 2008.

Chapter 7. The Pregnancy Paradox

"The beginning is glorious": Nora Ephron, *Heartburn* (New York: Knopf, 1983), p. 45.

a woman who has her first child before age twenty: National Cancer Institute Fact Sheet, "Reproductive History and Breast Cancer Risk," May 10, 2011, available at http://www.cancer.gov/cancer topics/factsheet/Risk/pregnancy.

in the 1930s the average age: Associated Press, "Age Increases for Motherhood," *St. Petersburg Times,* December 2, 1948.

Before 1960, nearly one-third of American females: "Teen Pregnancy," *Encyclopedia of Children and Childhood in History and Society,* 2008, at http://www.faqs.org/childhood/So-Th/Teen-Pregnancy.html.

Since 1970, the percentage of women: T. J. Mathews and Brady E. Hamilton,

"Delayed Childbearing: More Women Are Having Their First Child Later in Life," *National Center for Health and Statistics Data Brief,* no. 21 (August 2009), available at http://www.cdc.gov/nchs/data/databriefs/db21.pdf.

Lakshmanaswamy envisions a hormone patch: Raj Lakshmanaswamy, assistant professor of pathology, Texas Tech University Health Sciences Center, author interview, September 30, 2010.

developing a fake-pregnancy drug: As you might imagine, the idea of giving healthy women large doses of hormonal drugs to prevent a disease they might never get is still extremely controversial. Of course, millions of healthy women are already taking hormones — witness the pill. In another scheme, Malcolm Pike has been thinking for years about how to incorporate his cancer-protection ideas into a contraceptive device. Pike wants to essentially "fix" the pill so that it prevents breast cancer at the same time that it prevents pregnancy. "The amazing thing is that tens of millions of women took the pill this morning," he says. "If only one could get it right, one could get at the disease." Rather than a pill mimicking pregnancy in order to work, though, Pike envisions a better pill as more closely mimicking menopause. Currently, the pill stops ovulation, which is good for preventing cancer. But then the pill essentially replaces all those natural hormones, which is not good. Pike wants to stop ovulation by a different route, by blocking an upstream hormonal signal from the hypothalamus called gonadotropin-releasing hormone, or GnRH. Then he'd put back only a small amount of the lost estrogen and progesterone, just enough to keep women from feeling like crones, but not enough to stimulate breast or uterine cells. Problematically, GnRH is a peptide and would break down in the stomach if taken as a pill, so Pike has imagined this anticancer elixir as a nasal spray. He had the whole spritzing device packaged and ready to bring to market about ten years ago, but he couldn't get any investors or pharmaceutical companies to bite. "The best thing I ever did was to propose adding a third component to the pill," he now reminisces. "I still have to find a pharmaceutical company that would be willing to market it and that's a big *if*. I still think something like this is the way to go. Stop ovulation. Some way. It won't happen in my lifetime. We'll sort it out; it's just slow."

Janet Daling published results: Janet R. Daling et al., "Risk of Breast Cancer among Young Women: Relationship to Induced Abortion," *Journal of the National Cancer Institute,* vol. 86, no. 21 (1994), pp. 1584–1592.

Pro-life groups even sought legal action: Chinué Turner Richardson et al., "Misinformed Consent: The Medical Accuracy of State-Developed Abortion Counseling Materials," *Guttmacher Policy Review,* vol. 9, no. 4 (2006), pp. 6–11.

"was wonderfully rapid and its course excessively malignant": Samuel Weissel Gross, *A Practical Treatise on Tumors of the Mammary Gland* (New York: D. Appleton, 1880), p. 146.

A study in 2011 found that the more times a woman gives birth: Amanda I. Phipps et al., "Reproductive History and Oral Contraceptive Use in Relation to Risk of Triple-Negative Breast Cancer," *Journal of the National Cancer Institute,* vol. 103, no. 6 (2011), pp. 470–477.

In the United States, about 3,500 cases: Karen Hassey Dow, "Pregnancy and Breast Cancer," *Journal of Obstetric, Gynecologic, and Neonatal Nursing,* vol. 29, no. 6 (2000), pp. 634–640.

Chapter 8. What's for Dinner?

"First we nursed our babies": Mary McCarthy, *The Group* (New York: Harcourt, Brace, 1991), p. 291.

"crawl" to the nipple: Marshall Klaus has written a nice description of the breast crawl in the journal *Pediatrics:* "While moving up, he often turns his head from side to side. As he comes close to the nipple, he opens his mouth widely and, after several attempts, makes a perfect placement on the areola of the nipple." In addition to visual cues, scent also plays a major role in the crawl. If the right breast is washed with soap and water, the infant will crawl to the left breast and vice versa. Marshall Klaus, "Mother and Infant: Early Emotional Ties," *Pediatrics,* vol. 102, no. 5 (1998), pp. 1244–1246.

Breast-feeding may have helped the species evolve: Before the recent dawn of antibiotics, many women died shortly after childbirth, not only from puerperal fever (a post-hemorrhage infection of the genital tract), but also from "milk fever," or breast infections. To help prevent dangerously engorged breasts, sometimes small puppies were brought in to suck off the milk (I kid you not). Women also applied "suction cups" to each other. For more in this vein, read Valerie Fildes's excellent *Breasts, Bottles and Babies: A History of Infant Feeding* (Edinburgh: Edinburgh University Press, 1986).

"slight sleepiness, euphoria": Klaus, "Mother and Infant."

It was the late 1960s: Penny Van Esterik, "The Politics of Breastfeeding," in Stuart-Macadam and Dettwyler (eds.), *Breastfeeding,* p. 149.

Archaeologists have found four-thousand-year-old graves: Tina Cassidy, *Birth: The Surprising History of How We Are Born* (Boston: Beacon Press, 1999), p. 235.

the tight corsets of the Restoration: Fildes, *Breasts, Bottles and Babies,* p. 102.

"while taking every precaution": Plato, quoted in Naomi Baumslag and Dia Michels, *Milk, Money, and Madness: The Culture and Politics of Breastfeeding* (Westport, Conn.: Bergin & Garvey, 1995), p. 40.

"If a man has given his son to a wet-nurse": Quoted from C. H. W. Johns, *Babylonian and Assyrian Laws and Letters* (Edinburgh: T&T Clark, 1904), p. 61.

Even today, a child born in a developing country: Baumslag and Michels, *Milk,*

Money, and Madness, p. 8.

mortality rates reached 50 percent: Claire Tomalin, *Jane Austen: A Life* (New York: Knopf, 1997), pp. 7–9.

Jane Austin's story: Tomalin, *Jane Austen,* pp. 7–9.

Some sources claim this is where the term *farmed out:* Baumslag and Michels, *Milk, Money, and Madness,* p. 46.

"the bread and butter": For this quote by John Keating and a good overview of the early days of pediatrics, see Rima D. Apple, *Mothers and Medicine: A Social History of Infant Feeding, 1890–1950* (Madison: University of Wisconsin Press, 1987), p. 55ff.

"good Swiss milk and bread": As quoted in Apple, *Mothers and Medicine,* p. 9.

rise of germ theory: For a good discussion of this and its influence in separating humans from nature, see Linda Nash's *Inescapable Ecologies: A History of Environment, Disease and Knowledge* (Berkeley: University of California Press, 2006).

mother was sent home with a pat on the back: "Infant Food, Nestle's Lactogen," National Museum of American History, at http://americanhistory.si.edu/collections/object.cfm?key=35&objkey=110 (accessed October 12, 2011).

"It just didn't seem fair": Marian Thompson, quoted in Margot Edwards and Mary Waldorf, *Reclaiming Birth: History and Heroines of American Childbirth Reform* (Trumansburg, N.Y.: Crossing Press, 1984), p. 88.

"You didn't mention 'breast' in print": Edwina Froehlich, quoted in Emily Bazelon, "Founding Mothers: Edwina Froehlich, b. 1915," *New York Times,* December 23, 2008.

In Ghana, only 4 percent of women: Laurie Nommsen-Rivers, research assistant professor, Cincinnati Children's Hospital Medical Center, author interview, October 2010. For Sacramento rates, see Nommsen-Rivers, "Delayed Onset of Lactogenesis among First-Time Mothers Is Related to Maternal Obesity and Factors Associated with Ineffective Breastfeeding," *Journal of Clinical Nutrition,* vol. 92, no. 3 (2010), pp. 574–584.

the activists asserting we're in the midst of "a biocultural crisis": Dettwyler, "Beauty and the Breast," p. ix.

"And in any case, if a breast-feeding mother": Hanna Rosin, "The Case against Breast-Feeding," *Atlantic,* April 2009, accessed online at http://www.theatlantic.com/magazine/archive/

2009/04/the-case-against-breast-feeding/7311/.

formula confers the same average loss in points: Herbert L. Needleman et al., "Deficits in Psychological and Classroom Performance of Children with Elevated Dentine Lead Levels," *New England Journal of Medicine,* vol. 300, no. 3 (1970), pp. 679–695.

Two major reviews of the literature: Christopher G. Owen et al., "Effect of Infant Feeding on the Risk of Obesity across the Life Course: A Quantitative Review of Published Evidence," *Pediatrics,* vol. 115, no. 5 (2005), pp. 1367–1377; and S. Arenz, "Breast-Feeding and Childhood Obesity — A Systematic Review," *International Journal of Obesity,* vol. 28 (2004), pp. 1247–1256.

Chapter 9. Holy Crap

"O, thou with the beautiful face": Susruta Samhita, quoted in Valerie Fildes, *Breast, Bottles and Babies: A History of Infant Feeding* (Edinburgh: Edinburgh University Press, 1986), p. 14.

equivalent of one thousand light trucks: Daniel W. Sellen, "Evolution of Infant and Young Child Feeding: Implications for Contemporary Public Health," *Annual*

Review of Nutrition, vol. 27 (2007), pp. 123–148. Also, see A. M. Prentice and Ann Prentice, "Energy Costs of Lactation," *Annual Review of Nutrition,* vol. 8 (1988), pp. 63–79.

"Breastfeeding is a form of matrotropy": Sandra Steingraber, *Having Faith: An Ecologist's Journey to Motherhood* (Cambridge, Mass.: Perseus, 2001), p. 214. (Note, the more conventional spelling for this is *matrotrophy.*)

In the old days, people used to measure milk output: In fact, some doctors told women the only way to be absolutely certain their babies were getting enough to eat was to weigh them before and after *every feed,* including the two-in-the-morning one. It was another highly effective incentive to switch to formula.

In a paper describing the work: Jacqueline C. Kent et al., "Breast Volume and Milk Production during Extended Lactation in Women," *Experimental Physiology,* vol. 84 (1999), pp. 435–447.

In addition to recruiting the good bugs, these sugars prevent the bad bugs: A quick word about the use here of "good" and "bad" bacteria. As scientists learn more about the role of microflora in our bodies, these terms appear somewhat reductive because what the scientists really mean is the overall healthful balance of bugs. I continue to use these terms, though, because that is the way people described them to me and it still seems apt enough when talking about the role of milk sugars and microbes.

a dreadful disease called NEC: Information on NEC and premature babies from Lars Bode, assistant professor of pediatrics, University of California, San Diego, author interview, October 2010.

"We'll take little tiny droplets of milk": Video by CBS/Smartplanet.com, August 26, 2010, can be accessed at http://www.smartplanet.com/video/is-the-cure-for-cancer-inside-milk/460136.

There are ten times more microbacteria in our guts: Roderick I. Mackie et al., "Developmental Microbial Ecology of the Neonatal Gastrointestinal Tract," *American Journal of Clinical Nutrition,* vol. 69, no. 5 (1999), pp. 1035S–1045S.

Nearly a billion people don't live near clean drinking water: World Health Organization, "Global Health Observatory: Use of Improved Drinking Water Sources," available at http://www.who.int/gho/mdg/environmental_sustainability/situation_trends_water/en/index.html (accessed October 2011).

One Japanese company: For information on this and other products being developed with lactoferrin and marketed, and the economic analysis, see Vadim V. Sumbayev et al. (eds.), *Proceedings of the World Medical Conference: Malta, September 15–17, 2010* (Stevens Point, Wisc.: WSEAS Press, 2010), available at http://www.wseas.us/e-library/conferences/2010/Malta/MEDICAL/MEDICAL-00.pdf.

HAMLET kills forty different types of cancer cells in a dish: Catharina Svanborg et al., "Hamlet Kills Tumor Cells by an Apoptosis-like Mechanism — Cellular, Molecular and Therapeutic Aspects," *Advances in Cancer Research,* vol. 88 (2003), pp. 1–29.

several studies found: For example, see X. O. Shu et al., "Breastfeeding and Risk of Childhood Acute Leukemia," *Journal of the National Cancer Institute,* vol. 91, no. 20 (1999), pp. 1765–1772; for a more recent (and somewhat less enthusiastic) review of this literature, see Jeanne-Marie Guise et al., "Review of Case-Control Studies Related to Breastfeeding and Risk of Childhood Leukemia," *Pediatrics,* vol. 116, no. 5 (2005), pp. e724–e731.

donor milk is used mostly in neonatal

intensive care units: For an interesting discussion of markets for breast milk, see Linda C. Fentiman, "Marketing Mothers' Milk: The Commodification of Breastfeeding and the New Markets in Human Milk and Infant Formula," *Nevada Law Journal* (2009), available at Pace Law Faculty Publications, Paper 566: http:// digitalcommons.pace.edu/lawfaculty/566.

mothers of sons produced fatter, more- energy-dense milk: Katherine Hinde, "Richer Milk for Sons but More Milk for Daughters: Sex-Biased Investment during Lactation Varies with Maternal Life History in Rhesus Macaques," *American Journal of Human Biology,* vol. 21, no. 4 (2009), pp. 512–519. Also, Katherine Hinde, author interview, December 2010.

Babies have evolved their own tricks: David Haig, "Genetic Conflicts in Human Pregnancy," *Quarterly Review of Biology,* vol. 68, no. 4 (December 1993), pp. 495– 532; Sarah Blaffer Hrdy, *Mother Nature: Maternal Instincts and How They Shape the Human Species* (New York: Ballantine Books, 1999), pp. 430–441.

When the baby is older than one year: Dror Mandel et al., "Fat and Energy Contents of Expressed Human Breast Milk in Prolonged Lactation," *Pediatrics,* vol. 116, no. 3 (2005), pp. e432–e435.

After the terrorist attacks of 9/11: Sandra Steingraber, *Raising Elijah* (Philadelphia: Da Capo Press, 2011), p. 19.

A mother loses up to 6 percent of her calcium: J. M. Lopez, "Bone Turnover and Density in Healthy Women during Breastfeeding and after Weaning," *Osteoporosis International,* vol. 6, no. 2 (1996), pp. 153–159.

"I was storing some of my milk": Eleanor "Bimla" Schwarz, assistant professor of medicine, epidemiology, obstetrics, gynecology, and reproductive sciences, University of Pittsburgh, author interview, October 2010.

humans are the only primates: Daniel W. Sellen, "Evolution of Infant and Young Child Feeding: Implications for Contemporary Public Health," *Annual Review of Nutrition,* vol. 27 (2007), pp. 123–148.

"a pattern known to be optimal": Daniel W. Sellen, Canada Research Chair in Human Ecology and Public Health Nutrition, University of Toronto, author interview, October 2010.

a recent paper of his: Sellen, "Evolution and Infant Young Child Feeding."

the very stuff itself is oddly compromised: For an interesting discussion of how the profile of milk fats has changed due to the omega-6–dominant Western diet, see Erin E. Mosley, Anne L. Wright, Michelle K. McGuire, and Mark A. McGuire, "*Trans* Fatty Acids in Milk Produced by Women in the United States," *American Journal of Clinical Nutrition,* vol. 82, no. 6 (2005), pp. 1292–1297.

Chapter 10. Sour Milk

"To recognize milk which is bad": Ebers Papyrus, quoted in Valerie Fildes, *Breasts, Bottles and Babies: A History of Infant Feeding* (Edinburgh: Edinburgh University Press, 1986), p. 5.

"Today, polyurethanes can be found in virtually everything": "History," Centers for the Polyurethanes Industry, Polyeurethane.org, available at http://www.polyurethane.org/s_api/sec.asp?cid=853&did=3487 (accessed October 2011).

A typical home filled with polyurethane products": Bob Luedeka, executive director, Polyurethane Foam Association, author interview, August 2011.

It's questionable whether or not these substances: Y. Babrauskas et al., "Flame Retardants in Furniture Foam: Benefits and Risks," *Fire Safety Science Proceedings, 10th International Symposium, Interna-*

tional *Association for Fire Safety Science* (2011, pending publication).

Most deaths in fires are caused: Centers for Disease Control and Prevention, "Fire Deaths and Injuries: Fact Sheet," October 1, 2010, available at http://www.cdc.gov/homeandrecreationalsafety/fire-prevention/fires-factsheet.html.

They accumulate more toxins than other organs: France P. Labreche, "Exposure to Organic Solvents and Breast Cancer in Women: A Hypothesis," *American Journal of Industrial Medicine,* vol. 32 (1997), pp. 1–14.

Morton Biskind examined a pregnant woman: Morton Biskind et al., "DDT Poisoning: A New Syndrome with Neuropsychiatric Manifestations," *American Journal of Psychotherapy,* vol. 3, no. 2 (1949), pp. 261–270; Morton S. Biskind, "Statement on Clinical Intoxication from DDT and Other New Insecticides," presented before the Select Committee to Investigate the Use of Chemicals in Food Products, United States House of Representatives, December 12, 1950, Westport, Conn., published in the *Journal of Insurance Medicine,* vol. 6, no. 2 (March–May 1951), pp. 5–12.

finding DDT in the milk: For a great

overview of the problem, see Steingraber, *Having Faith,* p. 252. The 1951 study by E. P. Laug is recounted in "DDT and Its Derivatives," published by the United Nations Environmental Programme and the World Health Organization in 1979, available at http://www.inchem.org/documents/ehc/ehc/ehc009.htm.

Researchers in the Great Lakes region: one of the most famous studies is Joseph L. Jacobson and Sandra W. Jacobson, "Intellectual Impairment in Children Exposed to Polychlorinated Biphenyls in Utero," *New England Journal of Medicine,* vol. 335 (1996), pp. 783–789.

animal poisoning in Michigan in 1974: Åke Bergman, "The Abysmal Failure of Preventing Human and Environmental Exposure to Persistent Brominated Flame Retardants: A Brief Historical Review of BRFs," in Mehran Alaee et al. (eds.), *Commemorating 25 Years of Dioxin Symposia* (Toronto: Twenty-fifth Dioxin Committee, 2005), pp. 32–40. Also see Joyce Egginton, *The Poisoning of Michigan* (East Lansing: Michigan State University Press, 1980).

In the years that followed: For the long-term effects in the people exposed in the Michigan case, see Heidi Michels Blanck et al., "Age at Menarche and Tanner Stage in Girls Exposed in Utero and Postnatally to Polybrominated Biphenyl," *Epidemiology,* vol. 11, no. 6 (2000), pp. 641–671. Also, Michele Marcus, Pediatric Environmental Health Specialty Unit, Rollins School of Public Health, Emory University, author interview, November 2010.

at 36 parts per billion: Arnold Schecter, professor of environmental and occupational health sciences, University of Texas School of Public Health, author interview, September 2004.

In humans as well as rodents: For example, see Ami R. Zota, "Polybrominated Diphenyl Ethers (PBDEs), Hydroxylated PBDEs (OH-PBDEs), and Measures of Thyroid Function in Second Trimester Pregnant Women in California," *Environmental Science and Technology,* published online, August 10, 2011.

In 2010, researchers in New York found: Julie Herbstman, "Prenatal Exposure to PBDEs and Neurodevelopment," *Environmental Health Perspectives,* vol. 118, no. 5 (May 2010), pp. 712–719.

A recent study found that California women: Kim G. Harley et al., "PBDE Concentrations in Women's Serum and Fecundability," *Environmental Health Per-*

spectives, vol. 118, no. 5 (May 2010), pp. 699–704.

A Danish study showed an association: Katharina Maria Main et al., "Flame Retardants in Placenta and Breast Milk and Cryptorchidism in Newborn Boys," *Environmental Health Perspectives,* vol. 115, no. 10 (October 2007), pp. 1519–1526.

breast-fed infants and toddlers have considerably higher levels: Daniel Carrizo et al., "Influence of Breastfeeding in the Accumulation of Polybromodiphenyl Ethers during the First Years of Child Growth," *Environmental Science and Technology,* vol. 41, no. 14 (2007), pp. 4907–4912.

Some studies have found that breast-fed babies: For example, see Jacobson and Jacobson, "Intellectual Impairment in Children Exposed to Polychlorinated Biphenyls in Utero," pp. 783–789.

Recent studies show that lactating mothers off-load: Kim Hooper et al., "Depuration of Polybrominated Diphenyl Ethers (PBDEs) and Polychlorinated Biphenyls (PCBs) in Breast Milk from California First-Time Mothers (Primiparae)," *Environmental Health Perspectives,* vol. 115, no. 9 (September 2007), pp. 1271–1275.

For other chemicals, the dump rate is even higher: As cited in Hooper et al., "Depuration of Polybrominated Diphenyl Ethers."

Mothers who breast-feed for a year: Cathrine Thomsen et al., "Changes in Concentrations of Perfluorinated Compounds, Poly brominated Diphenyl Ethers, and Polychlorinated Biphenyls in Norwegian Breast-Milk during Twelve Months of Lactation," *Environmental Science and Technology,* vol. 44, no. 24 (2010), pp. 9550–9556.

"likely to be carcinogenic in humans": Kyle Steenland et al., "Epidemiologic Evidence on the Health Effects of Perfluorooctanoic Acid (PFOA)," *Environmental Health Perspectives,* vol. 118, no. 8 (August 2010), pp. 1100–1108.

Adult female striped and bottlenose dolphins are actually the "purest": Jennifer E. Yordy et al., "Life History as a Source of Variation for Persistent Organic Pollutant (POP) Patterns in a Community of Common Bottlenose Dolphins (*Tursiops truncatus*) Resident to Sarasota Bay, FL," *Science of the Total Environment,* vol. 408, no. 9 (2010), pp. 2163–2172.

I decided to start with my house dust:

Chemical analysis of my house dust was done by Heather Stapleton, assistant professor of environmental chemistry, Nicholas School of the Environment and Earth Sciences, Duke University; author interview, December 2010.

seventy-six new and suspect flame-retardants: Jacob de Boer, "Editorial: Special Issue: Contaminants in Food — Brominated Flame Retardants," *Molecular Nutrition and Food Research,* vol. 52, no. 2 (2008), pp. 185–186.

Chapter 11. An Unfamiliar Wilderness

"Brave new world": *American Heritage Dictionary of the English Language,* 4th ed., available at http://www.wordnik.com/words/brave%20new%20world (accessed October 2011).

"Walking today in an unfamiliar bio-chemical wilderness": James S. Olson, *Bathsheba's Breast: Women, Cancer, and History* (Baltimore: Johns Hopkins University Press, 2002), p. 240.

Globally, breast cancer is the leading cause: F. Kamangar et al., "Patterns of Cancer Incidence, Mortality, and Prevalence across Five Continents: Defining Priorities to Reduce Cancer Disparities in Different Geographic Regions of the

World," *Journal of Clinical Oncology,* vol. 24 (2006), pp. 2137–2150.

"omnis cellula e cellula": Rudolf Ludwig Karl Virchow, *Die Cellularpathologie in ihrer Begründung auf physiologische und pathologische Gewebelehre* (Berlin: A. Hirschwald, 1858).

Humans are just about the only free-ranging animal: Susan Love, clinical professor of surgery, David Geffen School of Medicine, University of California, Los Angeles, and president, Dr. Susan Love Research Foundation, author interview, March 2009.

Domestic pets, if not spayed, get it: Mel Greaves, *Cancer: The Evolutionary Legacy* (New York: Oxford University Press, 2001), p. 210.

Among the Kaingang women in Paraná, Brazil: This is according to Edimara Patrícia da Silva et al., "Exploring Breast Cancer Risk Factors in Kaingáng Women in the Faxinal Indigenous Territory, Paraná State, Brazil, 2008," *Cadernos de Saúde Pública,* vol. 25, no. 7 (2009), pp. 1493–1500.

One papyrus recommends applying a plaster: James V. Ricci, *The Genealogy of Gynaecology: History of the Development of Gynaecology through the Ages*

(Philadelphia: Blakiston, 1943), p. 20, as cited in Marilyn Yalom, *A History of the Breast* (New York: Knopf, 1997), p. 206.

Anne of Austria, the mother of King Louis XIV: Yalom, *History of the Breast,* p. 217.

De Morbis Artificum Diatriba: Bernardino Ramazzini, *De Morbis Artificum Diatriba* (London: Printed for Andrew Bell et al., 1705), pp. 122–123.

"Childless women get it / And men when they retire": W. H. Auden, "Miss Gee" (1937), published in *Another Time* (New York: Random House, 1940).

woman's tumor was sometimes a different size: Olson, *Bathsheba's Breast,* p. 77.

"I am satisfied that in the ovary": George Thomas Beatson, quoted in Olson, *Bathsheba's Breast,* p. 78.

breast cancer rates in the United States: For breast cancer statistics, see "SEER Stat Fact Sheets: Breast," at http://seer.cancer.gov/statfacts/html/breast.html.

he headed to Hiroshima: For an excellent description of Malcolm Pike in Hiroshima, see Malcolm Gladwell, "John Rock's Error," *New Yorker,* March 13, 2000, pp. 52–63.

As early as the 1930s, scientists: Olson,

Bathsheba's Breast, p. 178.

tufted-ear marmosets: See Sarah Blaffer Hrdy, *Mothers and Others: The Evolutionary Origins of Mutual Understanding* (Cambridge, Mass.: Belknap Press, 2009), pp. 92–97.

"Not in our wildest dreams": Carl Djerassi, *The Pill, Pygmy Chimps and Degas' Horse* (New York: Basic Books, 1992), p. 58.

"Estrogen is to cancer what fertilizer": Roy Hertz, quoted in Olson, *Bathsheba's Breast,* p. 178.

"Your problems are too complicated": Djerassi, *Pill, Pygmy Chimps and Degas' Horse,* p. 135.

to make human breast cancer cells grow faster: Brian E. Henderson et al., "Endogenous Hormones as a Major Factor in Human Cancer," *Cancer Research,* vol. 42 (1982), pp. 3232–3239.

progesterone is just as bad, and possibly worse: Sandra Haslam, a physiologist from Michigan State University, has been studying the nefarious effects of progesterone on mammary glands for years. "We've been pointing the finger at the wrong hormone all these years," she told me (author interview, July 2011).

They ovulate approximately one hundred times: Beverly I. Strassmann, "Menstrual Cycling and Breast Cancer: An Evolutionary Perspective," *Journal of Women's Health,* vol. 8, no. 2 (March 1999), pp. 193–202.

Today in America, nearly 20 percent of women: Jane Lawler Dye, "Fertility of American Women: 2006," *Current Population Reports,* U.S. Census Bureau, August 2008, available at www.census.gov/prod/2008pubs/p20-558.pdf.

a marketing article from the University of Southern California: Alfred Kildow, "The Dashing Malcolm Pike," *USC Health Magazine,* Summer 1996, available at http://www.usc.edu/hsc/info/pr/hmm/96summer/pike.html (accessed October 2011).

Captive tigers and lions also suffer: Greaves, *Cancer,* p. 210.

By 1992, Premarin: Kathryn Huang and Megan Van Aelstyn, presentation of a Notre Dame case study, "Hormone Replacement Therapy and Wyeth," available at http://www.awpagesociety.com/images/uploads/Wyeth-Powerpoint.ppt (accessed October 2011).

"I think of the menopause as a deficiency disease": Quoted in Jane E.

Brody, "Physicians' Views Unchanged on Use of Estrogen Therapy," *New York Times,* December 5, 1975.

"living decay": Robert Wilson, quoted in Gary Null and Barbara Seaman, *For Women Only* (Toronto: Seven Stories Press, 1999), p. 751.

women "rich in estrogen": Robert Wilson, from *Feminine Forever* (1966), as quoted in Olson, *Bathsheba's Breast,* p. 181.

anthropologist Sarah Blaffer Hrdy: On caloric requirements of children, see Hrdy, *Mothers and Others,* p. 31. For a discussion of the grandmother hypothesis, see pp. 241–243.

Estrogen, miracle hormone that it is: Karen J. Carlson, Stephanie A. Eisenstat, and Terra Ziporyn, *The New Harvard Guide to Women's Health* (Cambridge, Mass.: Belknap Press, 2004), p. 375.

Million Women Study: For general information on the Million Women Study, see http://www.millionwomenstudy.org/.

The main culprit appeared to be progesterone: The risk of estrogen alone to breast cancer is confusing and under discussion. While the Million Women Study found a 66 percent higher risk for women taking estrogen alone, a more recent study found that it was actually moderately protective against breast cancer in women with hysterectomies. It still, however, found an increased risk of stroke. See A. Z. LaCroix et al., "Health Outcomes after Stopping Conjugated Equine Estrogens among Postmenopausal Women with Prior Hysterectomy: A Randomized Controlled Trial," *Journal of the American Medical Association,* vol. 305 (2011), pp. 1305–1314.

Overall, hormone therapy in Britain caused: V. Beral et al., "Breast Cancer and Hormone-Replacement Therapy in the Million Women Study," *Lancet,* no. 362 (2003), pp. 419–427.

Chapter 12. The Few. The Proud. The Afflicted.

"Do unto those downstream": Wendell Berry, *Citizenship Papers* (Berkeley: Shoemaker & Hoard, 2003), p. 214.

Although the military knew: Agency for Toxic Substances and Disease Registry, "Analyses and Historical Reconstruction of Groundwater Flow, Contaminant Fate and Transport, and Distribution of Drinking Water within the Service Areas of the Hadnot Point and Holcomb Boulevard Water Treatment Plants and Vicinities,

U.S. Marine Corps Base Camp Lejeune, North Carolina, Chapter C: Occurrence of Selected Contaminants in Groundwater at Installation Restoration Program Sites," October 2010, p. C7.

At that time, analysis from one well: Agency for Toxic Substances and Disease Registry, "Analyses and Historical Reconstruction of Groundwater Flow, Contaminant Fate and Transport, and Distribution of Drinking Water," p. C94.

Tap water at the elementary school: Camp Lejeune Water System analysis document for dichloroethylene and trichloroethylene, North Carolina Department of Human Resources, Division of Health Services, Occupational Health Laboratory, February 4, 1985, analyzed and signed by John L. Neal.

TCE alone has been detected: Agency for Toxic Substances and Disease Registry, "Toxic Substances Portal — Trichloroethylene (TCE)," July 2003, at http://www.atsdr.cdc.gov/toxfaqs/tf.asp?id-172&tid-30.

present in 34 percent of the nation's drinking water: President's Cancer Panel, *Reducing Environmental Cancer Risk: What We Can Do Now, 2008–2009 Annual Report,* National Cancer Institute,

April 2010, p. 33, available at http:// deainfo.nci.nih.gov/advisory/pcp/annual Reports/index.htm.

In September 2011, the EPA formally reclassified TCE: For the EPA's assessment report, released September 29, 2011, see http://www.epa.gov/iris/supdocs/ 0199index.html.

is still used by most dry-cleaners: Ray Smith, "The New Dirt on Dry Cleaners," *Wall Street Journal,* July 28, 2011.

once used as an aftershave: Christopher Portier, director, Agency for Toxic Substances and Disease Registry, author interview, July 2011.

One recent European study: Sara Villeneuve et al., "Occupation and Occupational Exposure to Endocrine Disrupting Chemicals in Male Breast Cancer: A Case–Control Study in Europe," *Occupational and Environmental Medicine,* vol. 67, no. 12 (2010), pp. 837–844.

Vinyl chloride has been linked to breast cancer: Peter F. Infante et al., "A Historical Perspective of Some Occupationally Related Diseases in Women," *Journal of Occupational and Environmental Medicine,* vol. 36, no. 8 (1994), pp. 826–831. See also S. Villeneuve, "Breast Cancer Risk by Occupation and Industry: Analysis of the

CECILE Study, a Population-Based Case–Control Study in France," *American Journal of Industrial Medicine,* vol. 54, no. 7 (2011), pp. 499–509.

Another study found a very moderately increased risk: A. Blair et al., "Mortality and Cancer Incidence of Aircraft Maintenance Workers Exposed to Trichloroethylene and Other Organic Solvents and Chemicals: Extended Follow Up," *Journal of Occupational and Environmental Medicine,* vol. 55, no. 3 (1998), pp. 161–171.

Some studies found that dry-cleaning workers: P. R. Band et al., "Identification of Occupational Cancer Risks in British Columbia," *Journal of Occupational and Environmental Medicine,* vol. 42, no. 3 (2000), pp. 284–310.

other studies found a lower incidence: A. Blair et al., "Cancer and Other Causes of Death among a Cohort of Dry Cleaners," *British Journal of Industrial Medicine,* vol. 47, no. 3 (1990), pp. 162–168.

A 1999 study looking at Danish women: Johnni Hansen, "Breast Cancer Risk among Relatively Young Women Employed in Solvent-Using Industries," *American Journal of Industrial Medicine,* vol. 36, no. 1 (1999), pp. 43–47.

a set of studies looked at women on Cape Cod: Ann Aschengrau et al., "Perchloroethylene-Contaminated Drinking Water and the Risk of Breast Cancer: Additional Results from Cape Cod, Massachusetts, USA," *Environmental Health Perspectives,* vol. 111, no. 2 (February 2003), pp. 167–173.

The American Cancer Society attributes only 2 to 6 percent: Brett Israel, "How Many Cancers Are Caused by the Environment?" *Scientific American,* May 21, 2010, accessed at http://www.scientificamerican.com/article.cfm?id=how-many-cancers-are-caused-by-theenvironment; see also Elizabeth T.H. Fontham, "American Cancer Society Perspectives on Environmental Factors and Cancer," *CA: A Cancer Journal for Clinicians,* vol. 59, no. 6 (2009), pp. 343–351.

hot spots for breast cancer: J. Griffith et al., "Cancer Mortality in US Counties with Hazardous-Waste Sites and Ground-Water Pollution," *Archives of Environmental Health,* vol. 44 (1989), pp. 69–74.

report released in April 2010: President's Cancer Panel, *Reducing Environmental Cancer Risk.*

cancers caused by chemicals have been "grossly underestimated": Podcast interview with Margaret Kripke, professor of immunology and executive vice president and chief academic officer, University of Texas MD Anderson Cancer Center, February 7, 2011, available at http://www .commonweal.org/new-school/audiofiles/ podcast/97_m_kripke_final_w_intro.mp3.

Most of the major breast cancer organizations say: Denise Grady, "U.S. Panel Criticized as Overstating Cancer Risks," *New York Times,* May 6, 2010.

they account for little over half of all breast cancers: Hansen, "Breast Cancer Risk among Relatively Young Women."

Chapter 13. Are You Dense?

"Death in old age is inevitable": Richard Doll, quoted in Siddhartha Mukherjee, *The Emperor of Maladies* (New York: Scribner, 2010), p. 462. Original quote from Richard Peto et al., "Mortality from Smoking Worldwide," *British Medical Bulletin,* vol. 52, no. 1 (1996), pp. 12–21.

BRCA genes are most commonly found: Marie E. Wood, "A Clinician's Guide to Breast Cancer Risk Assessment," *Sexuality, Reproduction and Menopause,* vol. 8, no. 1 (2010), pp. 15–20.

Or my grandmothers could have inherited: Susan L. Neuhausen, "Founder Populations and Their Uses for Breast Cancer Genetics," *Breast Cancer Research,* vol. 2, no. 2 (2000), pp. 77–81.

In families with histories of breast and ovarian cancer: Greg Gibson, *It Takes a Genome: How a Clash between Our Genes and Modern Life Is Making Us Sick* (Upper Saddle River, N.J.: FT Press, 2009), p. 30.

The average lifetime risk of breast cancer: "Breast Cancer Risk Assessment Tool," National Cancer Institute, at http://www.cancer.gov/bcrisktool/RiskAssessment.aspx?current_age=42&age_at_menarche=10&age_at_first_live_birth=30&ever_had_biopsy=0&previous_biopsies=0&biopsy_with_hyperplasia=0&rela (accessed October 2011).

mammograms wouldn't miss 20 percent: National Cancer Institute Factsheet, "Mammograms," available at http://www.cancer.gov/cancertopics/factsheet/detection/mammograms, accessed October 2011.

Menopausal women taking hormone replacement therapy: Norman Boyd, "Mammographic Density and Breast Cancer Risk: Evaluation of a Novel

Method of Measuring Breast Tissue Volumes," *Cancer Epidemiology, Biomarkers and Prevention,* vol. 18, no. 6 (2009), pp. 1756–1762.

Some studies show that wine drinkers: C. M. Vachon et al., "Association of Diet and Mammographic Breast Density in the Minnesota Breast Cancer Family Cohort," *Cancer Epidemiology, Biomarkers and Prevention,* vol. 9, no. 2 (2000), pp. 151–160.

the equivalent of about two additional breast cancers per year: Denise Grady, "Breast Cancer Seen as Riskier with Hormones," *New York Times,* October 19, 2010; see also Peter B. Bach, "Postmenopausal Hormone Therapy and Breast Cancer: An Uncertain Trade-off," *Journal of the American Medical Association,* vol. 15, no. 304 (2010), pp. 1719–1720; and Rowan T. Chlebowski et al., "Breast Cancer in Postmenopausal Women after Hormone Therapy — Reply," *Journal of the American Medical Association,* vol. 5, no. 305 (2011), pp. 466–467.

"Traditional medicine and public health practices": Nancy Langston, *Toxic Bodies: Hormone Disruptors and the Legacy of DES* (New Haven, Conn.: Yale University Press, 2010), p. 149.

Mammograms might work pretty well:

Even this statement is open to debate. We assume it to be true, but a recent large study in Europe found little change in mortality in women who received regular mammograms and women who didn't, and these women were over fifty. Death rates over both categories had improved, but the researchers attributed the change to better treatment, not to better screening. See P. Autier et al., "Breast Cancer Mortality in Neighbouring European Countries with Different Levels of Screening but Similar Access to Treatment: Trend Analysis of WHO Mortality Database," *British Medical Journal,* published online, July 29, 2011, available at http://www.ncbi.nlm.nih.gov/pmc/articles/PMC3145837/.

U.S. Preventive Services Task Force: For the task force's recommendations, see U.S. Preventive Services Task Force, "Screening for Breast Cancer," December 2009, at http://www.ahrq.gov/clinic/USpstf/uspsbrca.htm.

in Colorado, fully one-third of all breast cancers: Lori Jensen, "A Local Look at Mammograms for Women under 50," *Boulder Daily Camera,* February 28, 2010.

It's a well-recognized fact that most breast cancers: The 2003 National

Health Interview Survey looked at 361 women diagnosed with breast cancer between 1980 and 2003. Results revealed that 57 percent found their cancers on their own, either by self-examination or by accident. M. Y. Roth et al., "Self-Detection Remains a Key Method of Breast Cancer Detection for U.S. Women," *Journal of Women's Health,* vol. 20, no. 8 (August 20, 2011), pp. 1135–1139.

women in China received inadequate training: Lee Wilke, associate professor and director, UW Breast Center, University of Wisconsin School of Medicine and Public Health, author interview, February 2010.

one in Canada, which did find: Anthony B. Miller et al., "Canadian National Breast Screening Study 2: 13-Year Results of a Randomized Trial in Women Aged 50–59 Years," *Journal of the National Cancer Institute,* vol. 92, no. 18 (2000), pp. 1490–1499.

A recent study from Duke University: Lee Wilke et al., "Breast Self-Examination: Defining a Cohort Still in Need," *Proceedings of American Society of Breast Surgeons* (2009).

BRCA genes make breast cells more sensitive: A. Broeks et al., "Identification of Women with an Increased Risk of Developing Radiation-Induced Breast Cancer: A Case Only Study," *Breast Cancer Research,* vol. 9 (2007), pp. 106–114.

Chapter 14. The Future of Breasts

"The world is too much with us": William Wordsworth, ca. 1806, from Jack Stillinger, ed., *Selected Poems and Prefaces by William Wordsworth* (Boston: Houghton Mifflin, 1965), p. 183.

Susan Love: Love likes to say, "We know how to cure breast cancer really well in a mouse. The problem is, we don't know much about how cancer works in women" (author interview, April 2009). In hopes of conducting more research with women and less with rodents, the Army of Women, a partnership between the Dr. Susan Love Research Foundation and Avon Foundation for Women, aims to enlist one million women study volunteers from diverse backgrounds. For more information, see www.armyofwomen.org.

The medical community is getting better: Heide Splete, "10-Year Breast Cancer Survival Rates Improve," *Internal Medicine News Digital Network,* September 29, 2010, available at http://www.internalmedicine

news.com/specialty-focus/women-s-health/
single-article-page/10-year-breast-cancer
-survival-rates-improve.html.

Yet surprisingly few national research dollars: Tiffany O'Callaghan, "The Prevention Agenda," *Nature,* vol. 471, no. 7339 (March 24, 2011), pp. s2–s4.

breast cancer will, on average, shave thirteen years off a woman's life: Tomas J. Aragon et al., "Calculating Expected Years of Life Lost for Assessing Local Ethnic Disparities in Causes of Premature Death," *BMC Public Health,* vol. 8 (2008), p. 116.

Decades ago, microbiologist-turned-humanist René Dubos argued: René Dubos, *Mirage of Health: Utopias, Progress, and Biological Change* (New York: Harper, 1959), pp. 29, 110–111.

PERMISSION CREDITS

Portions of this book appeared in different form in the *New York Times Magazine* as "Toxic Breast Milk" (from the *New York Times,* Jan. 9, 2005. All rights reserved. Used by permission and protected by the copyright laws of the United States. The printing, copying, redistribution, or retransmission of this content without express written permission is prohibited); in *Slate* as "My Ikea Couch Reeks of Chemicals" and "Younger Girls, Bigger Breasts"; and in *O, The Oprah Magazine* as "Do Breast Self-Exams Work? (And If Not, Why Do We Keep Doing Them?)."

Epigraph Credits
p. 29: Jayne Mansfield quote used by permission of CMG Brands. **p. 29:** Excerpt from *Master Breasts* by Francine Prose. Copyright © 1998 by Francine Prose. Reprinted with permission of the Denise

Photograph Credits
Introduction: Levi Brown
Chapter 1: Joe Shere / MPTV Images
Chapter 2: *Fate* magazine, November 2000 / R. Crumb

Chapter 3: Ivan Mateev / iStockphoto

Chapter 4: Frederick's of Hollywood, Inc.

Chapter 5: George Silk / Getty Images

Chapter 6: Gossard Lingerie

Chapter 7: Bettmann / Corbis Images

Chapter 8: Claire Reid Photography / Snugabell Mom & Baby Gear

Chapter 9: MOMILK by Julien Bertheir / www.julienberthier.org

Chapter 10: Mauricio Alejo

Chapter 11: "The Human Condition Medical Corset Project: Human Condition AaAa," by Sarah Kariko, 2010 / photo by Neil Dixon, Yankee Imaging

Chapter 12: Patricia Izzo/www.izzophotography.com

Chapter 13: Fox Photos / Hulton Archive / Getty Images

Chapter 14: Chia Evers

ABOUT THE AUTHOR

Florence Williams is a contributing editor of *Outside* magazine, and her articles and essays have been widely anthologized. *Breasts* was named a finalist for the 2011 Columbia/Nieman Lukas Work-in-Progress Award. Williams lives in Boulder, Colorado.